Pelican Books

Future Weather

After completing his Ph.D. in astrophysics at the University of Cambridge, John Gribbin worked for five years on the editorial staff of the journal *Nature*, chiefly responsible for the daily science report in *The Times*. He left in 1975 to join the Science Policy Research Unit of the University of Sussex, working in the 'futures' team on a study of the likely impact of climatic change on world food supplies. A book about the work of SPRU entitled *Future Worlds* was the main result of this three-year project. Since 1978, John Gribbin has been Physics Consultant to the *New Scientist*.

John Gribbin has published many books on subjects ranging from astronomy through geophysics to climatic change, as well as two novels. These include *White Holes* (1977); *Climatic Change* (editor and contributor, 1978); *Timewarps* (1979); a novel, *The Sixth Winter* (co-author with Douglas Orgill, 1979); and *Genesis: The Origins of Man and the Universe* (1981). His latest publications are both concerned with human origins and evolution, in terms of fact, in the form of *The Monkey Puzzle* (co-author with Jeremy Cherfas, 1982), and fiction under the title *Brother Esau* (co-author with Douglas Orgill, 1982). As well as being a consultant to *New Scientist*, John Gribbin contributes regularly to the 'Futures' section of the *Guardian* and broadcasts for the BBC World Service, British Forces Radio and, more occasionally, Radio Two. He was an adviser for a Thames Television documentary on the greenhouse effect and for TV South's science programme, 'The Real World'.

In 1974, John Gribbin received Britain's premier science-writing award, the National Award sponsored by Glaxo and administered by the Association of British Science Writers. Married with two sons, John Gribbin was born in 1946 and lives in Lewes.

John Gribbin

Future Weather

Carbon Dioxide, Climate and
the Greenhouse Effect

Penguin Books

Penguin Books Ltd, Harmondsworth, Middlesex, England
Penguin Books, 625 Madison Avenue, New York 10022, U.S.A.
Penguin Books Australia Ltd, Ringwood, Victoria, Australia
Penguin Books Canada Ltd, 2801 John Street, Markham, Ontario, Canada L3R 1B4
Penguin Books (N.Z.) Ltd, 182–190 Wairau Road, Auckland 10, New Zealand

First published in the U.S.A. by Delacorte Press 1982
Published in Pelican Books 1983

Parts of this book are based on material which
first appeared in a pamphlet published by Earthscan in 1981

Made and printed in Great Britain by
Richard Clay (The Chaucer Press) Ltd, Bungay, Suffolk
Set in Monophoto Baskerville

For Hubert Lamb

May you live in interesting times

Chinese curse

Contents

Part Two
The Global Greenhouse

Acknowledgements

This is a book about the greenhouse effect – the possibility that a build-up of carbon dioxide in the atmosphere may lead to a warming of the Earth in our lifetimes. But it is also a book about climatic change in general, since the impact of mankind's activities on the weather can only be seen in proper perspective against the background of natural changes in the climate of our planet. What little I know of this immense subject has been learned over more than a decade as a journalist and science writer reporting on the theme, initially from a position of great ignorance. I am especially grateful to Hubert Lamb, who has suffered my often naive questioning for much of that time, and who encouraged me to probe deeper into the mysteries, and to his colleagues at the University of East Anglia's Climatic Research Unit, especially Tom Wigley, the Unit's present Director, and Mick Kelly, for the input they have provided to many of the ideas expressed here. Among the many other experts whom I have called on for information from time to time, André Berger, Will Kellogg, Joe King, Kirill Kondratyev, George Kukla, Stephen Schneider and Goesta Wollin deserve special mention. I would never have been prodded into writing a book on the greenhouse effect, however, if it had not been for Jon Tinker at Earthscan, who asked me to prepare a briefing document on the subject, and his colleague Kath Adams, who ensured that the document appeared in print and that the Stockholm meeting it was intended to contribute to actually happened. Following that meeting, Bert Bolin's comments on the original document enabled me to correct some errors and helped to convince me that I now knew enough about this complex problem to justify putting my version of the story on the record in more permanent form. But this is far from being the last word on the matter, and I am uncomfortably aware of how ideas on the subject changed even during the year I was working on the book.

The greenhouse effect is, however, certainly important enough to justify a progress report that is accessible to the non-scientist who may be baffled by some of the claims and counter-claims made about the prospect of an imminent global warming. I hope this progress report will enable you to sort out the wheat from the chaff among those claims and counter-claims and to decide for yourself where the heart of the problem lies.

Introduction

We live in exciting times, and the mystery of the changing climate provides one of the most exciting current scientific puzzles. Concern about the vulnerability of a seemingly overpopulated world to the effects of vagaries of climate on harvests and food supply increased throughout the 1970s, stimulated by droughts and floods, severe winters and scorching summers, in Europe, North and South America, the USSR, Africa, India and China. The impression many non-specialists are left with, after reading the newspapers and watching TV, is that something has gone wrong with the weather, and that these are freak events; on a longer perspective, the experts tell us that it was the good growing years of the middle twentieth century, the decades of the 'green revolution', that were unusual, and that the kind of variability we have experienced in the 1970s is more normal. Either way, as population continues to increase, and as people in the hungry Third World strive to catch up, in material terms, with the developed world, the prospect of continuing extreme fluctuations of weather is increasingly disturbing, and there is a growing need for soundly-based advice on the future trends of climate – advice on which farmers, governments and aid agencies might act.

After several decades in which climatic studies were the poor relation in meteorology, with efforts concentrated on improved weather forecasting for a few days ahead, rather than for years or decades ahead, a combination of increasing awareness of the problem and new techniques for probing the mysteries of past climates has now begun to produce a coherent, consistent explanation of how and why climatic changes have occurred in the past. It is becoming possible to use the climatic record determined by firm physical evidence to shed new light on historical changes, ranging from the rise and fall of food prices on the stock market to the rise and fall of civilizations.

The next step should be to use this developing understanding of natural climatic changes to predict future trends. Ironically, however, just at the time when we have become aware of the need for such forecasting, and just at the time when we are beginning to understand natural climatic changes, our own activities have introduced a new factor into the equation. By burning fossil fuel (oil and, especially, coal), by slash-and-burn agriculture, and by tearing down tropical rain forest, human beings are producing a rapid build-up of carbon dioxide in the atmosphere. This undoubtedly acts as a blanket around the Earth, trapping heat that would otherwise radiate away into space and causing the surface of our planet to warm up. Many climatologists believe that this 'greenhouse effect' could operate so efficiently that before the end of the present century – in less than twenty years' time – it will begin to dominate the natural fluctuations of climate, bringing about conditions unique in the long history of civilization. A minority view is that, while the greenhouse effect is certainly real, it may be much weaker and more easily coped with. Even a modest warming would, however, change the regional patterns of temperature and rainfall around the globe, disrupting established patterns of agriculture.

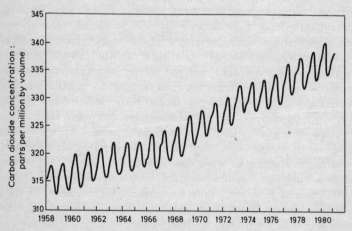

The build-up of atmospheric carbon dioxide recorded by monitoring instruments on Mauna Loa, Hawaii. As well as the regular seasonal fluctuation, there is a clear increase from the baseline level of 315 parts per million in 1958 to 338 ppm today. In 1959, the increase was 0·66 ppm; in 1981, it reached 1·81 ppm in a single year.

In the past, climatic changes affected human societies regarded as more primitive than our own. Are we more or less at risk than they were? On one level, we seem more able to cope with a changing climate, because we have better houses, warmer clothes, air conditioning, readily available energy and a more sophisticated technology. But on a more fundamental level – even leaving aside the question of how long we can rely on cheap energy – we are much worse off than our forebears. A hundred years ago, a nomadic tribe in the Sahel region of Africa could move south if the rain belts moved south. Today, the descendants of those people are fixed by political barriers to one country on the fringe of the desert. They cannot move south, because the lands to the south are already inhabited; growing population has meant a less flexible response to climatic shifts.

Before the second half of this century there was always a surplus of fertile land available, and room for people to move away from regions hit by drought, cold or other climatic extremes. Today the pattern is reversed. Migration is no longer a viable response to famine, and instead we move food around the world from the most productive regions (notably the grain belt of North America) to feed the hungry, either through the workings of the world food 'market' or as food aid.

This is the background to the present 'interesting times' in climatic studies. The need for reliable forecasts is clear, and the necessary understanding of natural processes is there. But the ironic uncertainty in forecasting arises from side effects of the main reason – population growth – that climatic forecasts are needed!

In this book I attempt to tackle the problem by first spelling out the impacts climatic changes have had in the past, the natural causes of climatic change, and the direction climate would be likely to take in the immediate future if left to its own devices. From this springboard it is then possible to jump off into the greater uncertainties of the greenhouse-effect debate and examine the evidence that the climatic applecart is about to be overturned by human activities. There are no certainties in any of this – climatologist Stephen Schneider is fond of the quip that 'the only certainties in life are death and taxes'. But there are very clear indications of the sorts of preparations that ought to be made as insurance against disaster; the kind of puzzles that have to be resolved are highlighted by the speed with which members of the pro-nuclear-power lobby have become 'environmentalists', expressing concern about the damage to the environment caused by carbon dioxide from coal-fired power stations. But the problems are not just local ones for individual countries or even for the 'Rich North' of our planet; suggestions that the Third World might actually *benefit* from an increase in atmospheric carbon dioxide, while we in the developed world suffer, introduce a new dimension into global politics and make an understanding at least of the outlines of the carbon dioxide greenhouse effect essential for anyone who cares about the future of humanity. I hope this book provides that outline.

John Gribbin
October 1981

Postscript

Since this book was written, progress in the study of climatic change has continued and the weather machine itself has thrown another severe winter, that of 1981–2, our way. The book is intended as an interim report on the present state of play, rather than a definitive last word on the greenhouse effect, but one new piece of work in particular should at least be mentioned here and now. Ronald Gilliland, who works at the High Altitude Observatory in Boulder, Colorado, has produced a series of scientific papers over the past year establishing beyond reasonable doubt that the Sun itself varies slightly in size on a time-scale of decades. As well as 'breathing' in and out – by a tiny amount – with regular rhythms of 11 and 76 years, the Sun's diameter has shrunk more steadily over the past 250 years, presumably as part of a longer-term cycle. Gilliland links these changes with some of the climatic changes on Earth in the past 250 years, suggesting that when the Sun is larger the Earth is cooler.

These claims are directly relevant to the greenhouse effect, and especially the work of Sherwood Idso mentioned in Chapter 9. Like other researchers mentioned in this book, Gilliland has tried to mimic the observed pattern of recent temperature fluctuations by using a computer to add in all of the supposed effects from volcanic dust, carbon dioxide and the newly discovered solar variation. His best results come from a combination of all three factors, but with a carbon dioxide greenhouse effect much more in line with Idso's calculations than those of other climate modellers.

As yet this proves nothing – except that intriguing new discoveries continue to be made, and that the investigation of the causes of climatic change is still far from over. But I hope this book will provide a background which will help you to understand and appreciate the rest of the saga of future weather as it unfolds before you.

John Gribbin
March 1982

Part One

The Changing Climate

1
Interesting Times

The 1970s brought a sequence of 'unusual' weather conditions around the world. Drought struck the Sahel region of Africa from 1968 to 1972, and again from 1975 onwards after only a year or two of 'normal' rainfall. Failure of the monsoons repeatedly brought problems for India, with the worst disaster in 1974; severe frosts wiped out the coffee crop in two Brazilian states in July 1975, with prices leaping on the world market as a result. In Europe, droughts in 1975 and 1976 were coupled with heavy, unseasonable rains in the Soviet grain belt and followed by severe winter weather in the late 1970s. And in the United States the story of the second half of the 1970s has been one of a succession of bitter winters (two 'worst in a hundred years' winters in succession, in 1976-7 and 1977-8) and severe droughts, affecting first one region and then another. It is natural to ask whether something has gone 'wrong' with our weather, and the question has been posed repeatedly in newspaper headlines and TV specials.

It is certainly true that these extreme fluctuations of weather followed a run of decades in which our climate had been relatively stable and well-behaved. Those were the decades of the green revolution, when it seemed that human technology had mastered the environment and agriculture was no longer at the mercy of wind and weather. But the return of less equable weather in the 1970s highlighted the fallacy of that complacent belief and showed humanity to be in many ways more vulnerable than ever before to the vagaries of natural shifts in the climatic balance.

In a keynote address to the World Climate Conference, called in 1979 by the World Meteorological Organization in the wake of growing concern about climatic change, Robert M. White of the US National Academy of Sciences stressed that recent climatic events are not unusual when placed in the right perspective. 'Similar events have occurred frequently in the historical record,' he said. 'What is new is the realization that vulnerability of human society to climatic events has not disappeared with technological development.' The point is that, at a time when population is still growing rapidly and food reserves are inadequate, we are vulnerable as never before, not so much to slow, long-term shifts in the average pattern of climate, but to erratic, unpredictable fluctuations, deviations from the long-term average, which occur from year to year and affect different regions of the globe in different ways. So it makes sense to look in a little more detail at those events of the recent past that have brought about this new realization of mankind's continuing vulnerability to climatic events before going on to look at the longer historical perspective and at the reasons for climatic fluctuations like those which still put so many millions of people at risk. In any such survey of the 1970s, the events in the Sahel region of Africa come first, both chronologically and because it was the succession of droughts in this part of the world that first brought home the message of vulnerability that Robert White summed up so succinctly in 1979.

The Sahel is the region of Africa immediately to the south of the Sahara desert which shares its name; strictly speaking, the Sahel extends across six African states, all of them in their modern form the product of the break-up of the old French empire in North Africa. These states of the Sahel proper are Mauritania, Senegal, Mali, Upper Volta, Niger (not to be confused with Nigeria, further south) and Chad; but the drought-affected region extends further east across Sudan and Ethiopia, making a band across Africa roughly in the latitude range from 10° to 20° north of the equator (Fig. 1.1). When climatologists, rather than politicians, refer to the Sahel, as often as not they include the whole of the drought-risk region, eight impoverished African states. Droughts in this part of Africa killed more people in the 1970s than all the minor wars and guerrilla activity which now tends to capture the limelight in reports from the region, and in the early 1970s it was the droughts that made the headlines.

When the climate of the region changed for the worse in the late 1960s, it came as an unexpected shock after a decade or so of improved living conditions which had seemed a tribute to mankind's mastery of a harsh environment. Most of the people of the Sahel are nomads, wandering the semi-arid region south of the desert to find

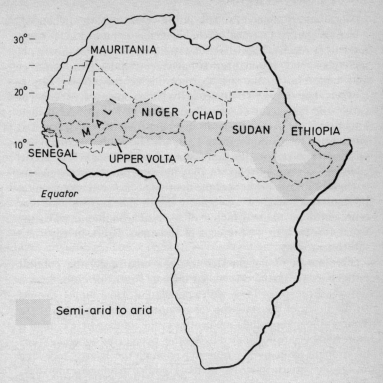

Figure 1.1. The drought-prone region of Africa stretches from the Sahel proper, south of the Sahara desert, across to Ethiopia in the east.

water and pasturage for their animals. Thus the division of the colonial empires into modern states with firmly drawn political boundaries was inevitably somewhat arbitrary, and, in cutting across traditional nomadic patterns of life, it always contained the seeds of conflict.

That did not seem to matter in the early and middle 1960s when foreign aid brought modern health care to the region, reducing the death rate and increasing the human population, at the same time that it paid for the drilling of deep wells, providing water for ever-increasing herds of cattle. To the nomads, cattle represent wealth as well as food; the herds grew and congregated around the wells, feeding off pasture maintained by the irrigation schemes and by a run of six years of good rainfall. But in 1968 the pattern changed. The rains that had seemed so reliable no longer pushed so far north towards the desert. The drought was nothing unusual for the region, merely an echo of droughts that have recurred down the centuries and came also in the early part of this century. Never before, however, had the Sahel built up such a large population of people or cattle; the herds first ate up all the grass and then turned their attention elsewhere, with cattle stripping trees of their leaves and goats grubbing up roots from the ground. The result was an ecological disaster, as only a little rain fell in 1969 and drought returned in 1970 and subsequent years. In 1975 the rains almost made a come-back, reaching 90 per cent of the figure that had seemed normal in the mid 1960s – but it was quickly shown that such rainfall now represented an unusual high, not 'normal' conditions, when the figure fell back to 30 per cent of the 1960s figure in 1976 and subsequent years.

The impact of the disaster was exacerbated by the political problems the colonial powers left behind. Niger, for example, is a state twice the size of Texas with a population of four million people. In the north of the state the inhabitants are chiefly nomads, the Tuareg; the main population centres in the south, where rainfall is more reliable, are dominated by another people, the Songhai. When the drought set in and the rains moved south the Tuareg, following an age-old pattern, moved south too. Huddled in refugee camps near the cities, they were afflicted by epidemics as well as hunger – and in 1973 Niger had only 82 trained doctors. Inevitably, the city-dwellers did not provide wholehearted support for the refugees, and in some cases there were inter-communal conflicts. Similar problems occurred between states in the Sahel, where governments saw themselves as in competition with other poor countries for food and other aid from the 'Rich North'. More than 100,000 people died in the Sahel proper as a result of the drought, while in the region to the east, around Sudan and Ethiopia, even more people may have died, although the reticence of governments in admitting their troubles makes it impossible to obtain accurate figures. The old Ethiopian regime was, indeed, overthrown partly as a consequence of troubles stemming from the drought. The deaths and injuries resulting from that civil war should perhaps also be laid at the door of the climatic changes.

The missing monsoon

Why did the rainfall move away to the south? The rains that the people of the Sahel depend on are monsoon rains. They come only at a certain time of year – in summer – and unless the climate shifts they can be relied on to come at the same time each year. Putting it at its simplest, the monsoon is brought by a westerly wind that moves off the sea carrying moist air over the land to the east. As the moist air rises over the land, it cools and drops its burden of moisture as rainfall. These winds are part of the overall pattern of circulation of the atmosphere, a pattern produced by the interplay of the effect of the Earth's rotation, the heating effect of the Sun which, by and large, makes warm tropical air rise and move away from the equator while colder air moves in from high latitudes to replace it, and the balance between land and sea. From season to season and year to year the oceans are less affected than continental land masses by changes in the amount of heat reaching them from the Sun, and are said to have a large thermal inertia. Compared with the land, the sea stays warm in winter and cool in summer, which is why maritime climates like those of Oregon or Britain show less extreme seasonal changes than continental climates like those of Illinois or Siberia.

Because of the way all these components of the weather machine interact, and especially because there is much more land in the Northern Hemisphere than in the Southern, the boundary between the northern and southern halves of the atmospheric circulation does not exactly match up with the geographical equator. In Northern Hemisphere summer the climatic equator – known as the Inter-tropical Convergence Zone, or ICZ – lies north of the equator; during northern winter it lies south of the equator. The summer westerly winds of the monsoon, so important to the Sahel as well as to India and Asia, are a feature of the atmospheric circulation south of the ICZ and north of the equator – they are in the Southern Hemisphere in climatic terms, but in the Northern Hemisphere geographically. That is just a quirk of the geography of our planet; what matters for the Sahel is that when the ICZ stays a little further to the south in summer, it keeps the monsoon rains down near the equator and prevents them reaching the semi-arid regions on the southern fringe of the Sahara desert.

So the question is really why the ICZ should fail to move as far northwards in summer now as it did in the early and middle 1960s. I shall look at the workings of the weather machine and atmospheric circulation in more detail in Chapter 3, but at its simplest the pattern can be thought of as one of concentric bands, each with its own characteristic pattern of wind flow and climate, centred on the pole in each hemisphere and extending to the equator, or more accurately to the ICZ. In each hemisphere the region over the pole itself is one of relative calm, surrounded by a dominating flow of westerly winds which sweep around the globe at high altitudes, the circumpolar vortex. This westerly flow is the mainstream of the atmospheric circulation, and changes in its pattern of behaviour affect the climate zones of an entire hemisphere. It exists as a result of the differences in heating of different latitude zones by the Sun and is only marginally influenced by the geographical distribution of land and sea, so it has definitely been a feature of planet Earth throughout most of its $4\frac{1}{2}$-thousand-million-year history. But those marginal influences of geography, including changes in the ice cover at high latitudes, and other effects which are the subject of the rest of the first part of this book, can be of crucial importance for us if, for example, they move the monsoon belt in Africa a few tens of miles to the south or bring drought to the great plains of North America. Without yet looking at why the circumpolar vortex might shift subtly from one decade to the next, it is clear that if the vortex expands away from the pole, then climatic zones, including the monsoon belt, are squeezed down towards the equator. Derek Winstanley, a climatologist based in Canada, and Reid Bryson, Professor of Meteorology and Geography at the University of Wisconsin–Madison, are among the specialists who have explained the disappearance of the rains from the Sahel as a side effect of just such an expansion of the circumpolar vortex in the Northern Hemisphere. The implication is that the problem is not 'just one of those things' affecting a few impoverished states in an economically unimportant part of the world, but part of a bigger pattern affecting at least the whole of the Northern Hemisphere. And Bryson, Winstanley and others who subscribe to this view have no difficulty in fitting the other Northern Hemisphere climatic disasters of the 1970s into this scheme of things.

The blocking highs

The great river of air that is called the circumpolar vortex follows a somewhat meandering path around the globe. It does not flow due east–west but zigzags slightly on its travels, weaving from north a little way south and back again from time to time. High above the ground, this wind blows so strongly that it is called the jet stream, and the flight paths of modern airliners flying across the Atlantic or Pacific are adjusted on flights from west to east to take advantage of the following wind, while on flights from east to west routes are chosen to avoid bucking the head wind. Leaving the tropics and the monsoon region for now, and taking the climatic story up to the latitudes of the jet stream, we are moving into the temperate latitudes of the rich, developed countries of North America and Europe. The jet stream itself does not flow at ground (or sea) level, but its influence is felt directly at the surface of the Earth in the form of the weather systems called depressions – low-pressure systems that sweep from west to east and follow, by and large, the track of the jet stream high above in the stratosphere. Where the jet stream zigs south, the depressions follow; where the jet stream zags north, there the depressions must surely go too. Depressions bring precipitation – rain or snow – and zigzags in the jet stream can bring warm air to higher latitudes, or chilling polar air diving southwards.

When the circumpolar vortex is strong, and the jet stream blows vigorously, it follows a relatively tight circle around the polar regions, with few zigzags. Climatic zones to the south have ample room to expand northwards, the monsoons occur with welcome regularity and the temperate regions have mild years with no great extremes of temperature or rainfall. But when the circumpolar vortex is weaker, so that the main flow of winds from west to east is more sluggish, then it zigzags more. The range of influence of the circumpolar vortex pushes further south, squeezing climatic zones towards the equator and restricting the northward penetration of the monsoon (Figure 1.2). At the same time, in the temperate zone which includes North America and Europe, the more sluggish jet stream and its accompanying surface-level depressions can be more easily deflected by other features of the atmospheric circulation system. Regions where the atmospheric pressure is high – anti-cyclones, or simply 'highs' – act like hills or islands of stability in the atmosphere, around which the depressions ('lows') and the jet stream flow. Unlike real mountains, of course, atmospheric highs are temporary features, and they also drift from west to east at these latitudes, but anyone can make a good stab at a weather forecast for twenty-four hours ahead by assuming that high-pressure systems

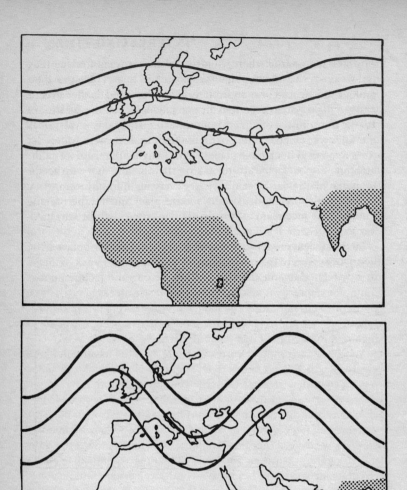

Figure 1.2. Changes in the circulation pattern of the Northern Hemisphere directly affect the monsoon rains of the Sahel. Strong west–east zonal circulation allows the rains to the south to push northwards; when the zonal circulation is weak and easily deflected the rains get pushed away to the south. The dotted area indicates schematically the extent of the monsoon in each case.

shown on the weather chart will drift a little eastwards, taking fine, dry weather with them, while the lows will scurry around their flanks, carrying rain or snow as they do so. Occasionally, however, a high-pressure system gets stuck in one place, for reasons which are still not fully understood. Such a blocking high may stay put for days or even weeks on end, acting like a real mountain around which the other weather systems have to move. Since highs are regions of calm, clear air, the land beneath them experiences fine, dry weather in summer and clear, cold weather with severe night frosts in winter. A summer high that stays too long in one place quickly becomes a problem, as a succession of fine, dry days may soon herald a drought (Figure 1.3).

All the evidence suggests that the climatic trials and tribulations of the 1970s and early 1980s – the 'interesting times' we live in – can be directly related to an expansion and weakening of the circumpolar

Figure 1.3. The weak circulation pattern also brings its problems further north. With strong circulation, rain-bearing depressions move regularly across Britain and into Europe. With weak circulation, a 'blocking high' may become established, diverting rain-bearing airstreams to produce drought, as happened in Britain in the summer of 1976, or bringing the depressions to a halt over Britain, producing heavy rainfall and flooding.

vortex, which has not only pushed the monsoon south but has allowed blocking highs to recur with unwelcome frequency, steering low pressure systems with their burdens of moisture on extreme zigzags north or south of their main west–east route around the world.

In 1972, a series of weather-related agricultural disasters around the world provided a classic example of the atmospheric 'teleconnections' that make the changing climate important for all of us, not just the nomads of the monsoon belt. In India, the failure of the monsoon in 1972 brought an 8-per cent drop in rice production; off Peru, the anchovy fishery which had been a basis of the region's economy failed; and droughts in Australia and South America emphasized that we are dealing with a global phenomenon, not one confined even to the Northern Hemisphere. The Sahel drought intensified; and Soviet food production fell by 8 per cent, with the region around Moscow experiencing its worst drought for 300 years. With floods in the mid-western United States, this was the worst year for agriculture worldwide in modern times, and total world food

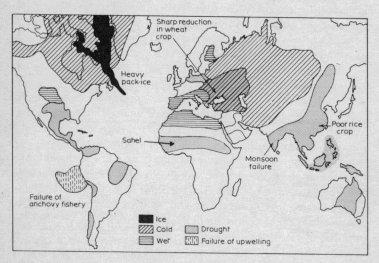

Figure 1.4. 1972 was a year of unusually bad weather worldwide. Drought and crop failures in five continents are detailed in the text. Some climatologists link these events with global cooling; but it is naive to assume that a global warming must therefore be a 'good thing'.

production fell by 2 per cent, the first fall since the Second World War. Nothing could bring home more clearly the fact that even with modern agricultural technology we cannot overcome the worst that nature can throw at us. As the Soviet Union made up its food deficit by purchasing 28 million tons of grain from North America, the price of meat and bread rose in American supermarkets, following the simple logic of supply and demand, and agriculture became a hot political issue. All this, however, was just a forerunner of things to come.

More freak weather

It was the turn of Europe to experience the next really dramatic run of so-called 'freak' weather. (The adjective may not be entirely appropriate, as we shall see later; what commentators mean is that the weather is unusual compared with conditions of the recent past. But is it unusual compared with the longer historical record – or the patterns we can expect in future?) Although the winter of 1974–5 was mild in England, in June 1975 snow fell on London, the first time this had happened at that time of year this century. Then, as if to confuse the public – and the crops – completely, the unseasonable chill gave way to the hottest July and August for many a year. Overall, 1975 turned out to be the fifth driest year of the twentieth century, thanks largely to blocking high conditions over western Europe, and as the rains stayed away through the winter the term drought began to be used. Further east, meanwhile, the Russians were having a bad time with their harvest again, and by the autumn of 1975 strong hints coming out of the USSR suggested a shortfall in grain even more severe than in 1972. From May 1975 to April 1976 was the driest twelve-month period on record in England and Wales, and the last thing anyone expected was that this would be followed by another scorching summer. Yet 1976 put even 1975 in the shade. The blocking high sat over England for week after week. After the hottest June of the century, following more than a year of drought, the River Thames itself dried up completely for nine miles at its source, while many reservoirs were down to one-fifth capacity. The French grain harvest was reduced by 25 per cent as all of western Europe baked in the sun. But to the east, again, it was a different story.

The blocking high that sat over Europe kept England, France and nearby regions dry, but the low-pressure systems, following the meandering path of the jet stream, still had to go somewhere with their burden of moisture, picked up on their passage over the Atlantic. Most of them went around the northern edge of the blocking high – the jet stream zigged northward – and then descended on the Baltic region around Moscow and Leningrad, dumping their rain there. Scotland, Scandinavia and the Baltic missed out on the scorching summer; too much rain proved as disastrous for the Soviet grain crop as too little did for the French, and once again the influence of the weather on agriculture disrupted economic planning on a global scale.

If you averaged out the rainfall across Europe and Russia, of course, there was no more or less than usual – it was just that one region had floods while another had drought. And if you averaged it out over the whole of the calendar year 1976, you got much the same misleading impression, for the great drought of 1975–6 was broken in England and Wales by torrential downpourings of rain in the autumn. The driest sixteen-month period in the meteorological record was followed by the second wettest September–October since records began. While the reservoirs filled up, puzzled farmers saw their cracked and parched fields turn into swamps.

While all this was going on in Europe, the great plains of the north-central United States had suffered one of the worst blizzards in the history of the region, from 10 January to 12 January 1975. The snows extended north into central Canada, and in the US covered North and South Dakota, Iowa, Nebraska and Minnesota, freezing more than 55,000 head of cattle to death and killing eighty people from exposure or heart attacks as they struggled against the blizzard. By late 1976, as the European drought washed away, it was the turn of North America to hit the international headlines once again, with almost exactly the same pattern of persistent climatic extremes that had been afflicting Europe – drought in the west, heavy precipitation in the east. The cause was the same – blocking high conditions and a zigzagging jet stream. But the effect was different, because the extreme conditions built up in winter, bringing not rain but snow to the north-east as low-pressure systems curved across Canada and dumped their moisture. The drought was far more important than the blizzards in economic terms, and it came right on cue in the sequence of droughts which have hit the great plains roughly every twenty years – including the notorious dust-bowl era of the 1930s –

since the beginning of the nineteenth century. More of this in due course; it was the snows in the north-east, however, that made the biggest splash in news coverage.

In the winter of 1976–7, millions of workers were laid off in the north-east, dozens of people died in the snow and the economy suffered a blow estimated at $6 billion, reducing the real Gross National Product of the whole US by one per cent, according to the US Chamber of Commerce Chief Economist, Jack Carlson. By mid-February 1977 the US Weather Service had declared officially that the winter was the worst, for the eastern two thirds of the country, since the founding of the Republic. Even Florida was hit by frost, which destroyed more than 10 per cent of the citrus crop, and the phrase 'worst for a hundred years' began to crop up regularly to describe one local climatic extreme after another. Many people took comfort from that phrase, assuming that after the 'worst winter for a hundred years' it would be a further hundred years before anything comparable happened again. Their naive optimism was shattered in January 1978, when another sequence of extreme weather conditions settled across the US. After a year-long drought, California was inundated by four weeks of steady rain and snow, with floods following the drought. In Louisville, Kentucky, 16 inches of snow fell; in Boston, Massachusetts, 21 inches fell in one blizzard in February 1978, and three weeks later a second storm dumped 27 inches of snow on the city in the space of 24 hours, accompanied by winds up to 100 mph in what was described as a 'snow hurricane'. Skiers were out in force, both there and in downtown New York, where a 'mere' 13 inches of snow fell in one blizzard on 20 January. How could two 'worst in a hundred years' winters follow in such rapid succession?

There are two answers to that question. The first is that just as a run of heads when tossing a perfectly balanced coin doesn't change the odds of 50:50 on the next toss coming up heads too, so the occurrence of one 'one in a hundred' winter (or drought) does not change the odds of 100:1 on the next winter being equally severe. But still, it would be very bad luck to get two such disasters in a row. The second answer, however, is more ominous. The occurrence of the 'worst winter for a hundred years' might mean it is a one-in-a-hundred freak. Or it might mean that the sort of winter that used to be *common* a century ago has returned to plague us, because of a shift in the natural climatic balance. The explanation of weather-related disasters in the 1970s in terms of a weakened atmospheric circulation, expanded circumpolar vortex and zigzagging jet stream exactly fits that interpretation of the evidence, and carries with it the grim prospect that such 'freak' weather conditions may remain with us for years or decades to come.

Into the eighties

Within this context, the European Economic Community initiated in 1978 a major study of climatic change and the way extremes of weather can affect our lives. When the decision was announced to the press, the EEC representatives drew attention to the remarkable sequence of extreme weather conditions Europe had experienced in the preceding fifteen years. These included the coldest winter since 1740, the driest winter since 1743, the mildest winter since 1834, the greatest drought since 1726 and the hottest month (July 1976) since records began three centuries ago. Over the same period, as we have seen, a similar pattern of extremes had been experienced world wide, although, of course, there are no records going back 300 years to provide a yardstick for measurements in North America and many other parts of the world. Following the EEC announcement, the pattern continued with an almost non-existent summer in much of Europe in 1978, followed by a spectacularly fine, dry, record-breaking autumn. In January 1979 Britain was hit by its worst winter on record, and the pattern of extremes has continued into the 1980s.

On 15 January 1981, the London *Times* carried a front-page story, under the headline 'Hundreds die as cold grips three continents', reporting the problems caused by severe snowstorms in Spain and south-west France, 'more than its accustomed share' of bad weather for Japan, and a state of emergency declared in Florida after two days of freezing weather which extensively damaged 'the state's multi-million dollar citrus, tomato and sugar cane crops'. A few weeks later, on 22 March, *The Times*'s sister paper the *Sunday Times* carried a more lengthy report on the severe problems being posed by flood and drought in different parts of China. Right up until December 1980 the People's Republic had prided itself on its ability to feed its 1,000 million people. But after a series of climate-induced agricultural disasters the Chinese government took the unprecedented step of calling in UN observers to take stock of the situation and formally requesting aid from the capitalist West to avoid famine. To some extent this reflects the fact that the present regime in China is more open than the regime of the 1960s and 1970s; to a far greater extent, however, it reflects the fact that this kind of agricultural disaster had not happened in China since the days of the civil war.

For Hubei province, some 500 miles south of Peking, the main problem was flooding which destroyed schools, hospitals, power stations, bridges, roads and 210,000 homes, causing damage estimated in economic terms as $1,000 million and the loss of 2½ million tons of grain.

For Hebei province, around Peking itself, the problem has been drought. In 1980, monthly rainfall figures for the province never went above 80 per cent of 'normal', and often fell as low as 30 per cent. Grain losses totalled 4½ million tons, and the underground water-level dropped so much that drinking supplies were affected. The problems faced by the population were then exacerbated by a bitter winter. As the *Sunday Times* put it, the scale of the problem can be put in perspective by the realization that Jingzhou, one prefecture within the province of Hubei, is the size of Holland. More than 43 million people were affected by the disasters in Hubei and Hebei provinces, and China reported a further 130 million people across five other provinces affected to a lesser extent by similar troubles. Since China has one third of the population of the entire developing world, and UN and other relief agencies are already stretched trying to cope with the problems afflicting the other 2,000 million members of the Third World, the prospect of China losing its proud self-sufficient status is a grim one for anyone concerned about the fate of our global society and the imbalance between the rich north and the poor south. Worse still, at a time when even China is seeking help from the rich north, the rich north has its own problems to contend with and may be less able than before to provide aid, especially food aid. The point was hammered home as drought returned to the US (it had scarcely been away!) in the late summer of 1980.

It will be no surprise by now to learn that the drought was caused by a blocking high over the western part of the US, a 'hill' of high pressure in the atmosphere which deflected rain-bearing systems northwards across Canada. Drought in the western US followed a dry summer and brought immediate problems; further east, blocking systems prevented moisture-bearing winds moving into the corn belt from the Gulf of Mexico, and after the tribulations of the 1970s the moisture content of the subsoil reached an 'all-time' low (which means the lowest value since people started bothering about such measurements in that part of the world, less than a hundred years ago).

A final irony was that the winds which 'should' have moved smoothly from west to east off the Pacific, dropping their moisture where it was needed and not getting in the way of the air-flow off the Gulf of Mexico into the grain belt, still had enough strength left to zig northwards again in the autumn and winter of 1980–81, blowing the storm track of depressions, which normally runs up the east coast and brings rainfall to New England, out into the Atlantic. Not one but three droughts, in the west, in the grain-belt and in the north-east, stemmed from one complex blocking system affecting the whole of North America (Figure 1.5). The science of meteorology is still unable to explain exactly why such a pattern of events should occur in one particular year, or to forecast the development of such a blocking system. But the science of climatology can put all these problems in their proper historical perspective. Just which years the

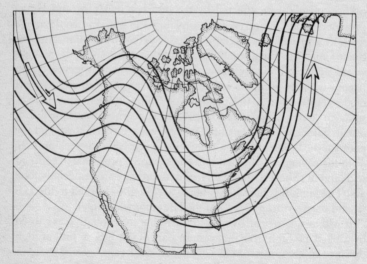

Figure 1.5. Blocking highs also disrupt the weather across North America. A persistent sustained kink in the flow of westerlies in the winter of 1980–81 brought a pattern of drought directly related to the pattern of the westerly winds, shown here for February 1981.

blocks will come in, and which regions will be affected by drought, we do not know; but we can now see that more frequent droughts and other climatic extremes are to be expected, since the weather of the world, far from bringing 'freak' conditions in the 1970s, is actually shifting back into a pattern that has dominated the globe for the past thousand years.

2
The Historical Perspective

The term 'normal weather' is meaningless without some explanation of the span of time over which the normal conditions have been determined. In most parts of the world the weather changes dramatically, although to a large extent predictably, in the course of a single year, with the 'normal' march of the seasons. But no two years are exactly alike, and no two decades exactly follow the same pattern of changing weather. The simplest definition of climate is indeed 'average weather' – but the average weather of the 1970s, say, was very different from the average weather of the 1670s, while only 20,000 years ago, a short time in the history of the Earth, our planet was in the grip of a full Ice Age. Climate is always changing, on all time scales; but this realization is very much a feature of modern science, for until well into the twentieth century meteorologists believed that the only changes in the weather were fluctuations around the average, and that given a long enough span of measurements to average they would be able to define the normal weather conditions – the climate – for any location on Earth.

It is hardly surprising that the variability of climate is a recent discovery of science, for it was only in 1840 that official records of temperature, rainfall and so on began to be kept, at the Royal Observatory, Greenwich, in London. At about the same time, the outspoken young President of the Swiss Society of Natural Sciences, Luis Agassiz, espoused the then controversial new theory that the scratched and polished boulders which were scattered across the Jura mountains had been left there by a great ice sheet which had melted thousands of years ago. The debate about the reality of the great Ice Age raged for decades, but even when, at the end of the nineteenth century, it had become an accepted fact that the Earth had experienced more than one Ice Age during its geological history, the scientific world pictured these as catastrophic events which had occurred in the remote past. The fact that the Earth could suffer a climatic change severe enough to cover large parts of North America and Europe with ice did not shake their faith that the climate today was essentially unchanging. That belief was only slowly undermined, as the growing meteorological record began to show directly that the weather of the mid twentieth century was different from the weather of the mid nineteenth century, and as improving geological techniques, coupled with historical studies, provided an ever clearer picture of the changing climate of historical times, especially in Europe.

Today there is such a comparative wealth of information about past climates that it is more important than ever to specify the period of time which is of particular interest. In terms of the whole history of the Earth – some 4½ thousand million years long – 'normal' weather is much milder than today, with ice-free polar regions and even what we think of as temperate latitudes covered with lush 'tropical' vegetation. Occasionally, because of the constant slow drift of the continents around the globe, geographical conditions occur which allow great ice sheets to build up over the poles, and then the Earth is plunged into an Ice Epoch which may last for a few million years or a few tens of millions of years. Within the Ice Epoch, ice sheets ebb and flow to produce a rhythmically changing pattern of full Ice Ages separated by short-lived interglacials, warmer than the full Ice Ages but nowhere near as warm as the long-term normal state of the

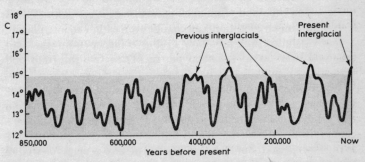

Figure 2.1. All of the period of warmth during which human civilization has developed and flourished represents a short-lived interglacial, a rare departure from the much colder conditions which have prevailed on Earth for millions of years. A variety of geological techniques reveals the changing level of global mean temperature in this figure compiled by the US National Academy of Sciences.

Earth. We live in such an interglacial, a temporary warm spell in between the Ice Ages, which are themselves part of an Ice Epoch that has gripped the Earth for more than 3 million years (Figure 2.1).

The present interglacial

Although on a time-scale of hundreds or thousands of millions of years the normal climate is warmer than today, on a time-scale of the past 2 million years the normal climate is much colder than today. I shall look in more detail at the causes of these changes, and what they tell us about the imminence of the next Ice Age, the end of our present interglacial, in Chapter 4. But in a sense those Victorian meteorologists who disregarded the ebb and flow of Ice Ages in their assessment of normal weather today were right – such a dramatic change is certainly on a different scale from anything that has happened in historical times, and the best guide to the kind of changes we might expect from decade to decade and century to century comes not from looking at Ice Age rhythms but from study of the way the weather of the world has changed within the present interglacial, since the end of the most recent Ice Age. That interglacial began just over 10,000 years ago, and it is no coincidence that the rise of human civilization on Earth has taken place over just that same 10,000-year span, jumping off from the revolutionary invention of agriculture in the millennia when ice was retreating and farming became a practicable proposition around the Mediterranean. Ten thousand years of climatic history is enough to give a feel for how much conditions can change even within an interglacial; a more detailed interpretation of the evidence from the past thousand years or so will then put in perspective the trials and tribulations caused by weather variability in our own century.

Unravelling the complexities of climatic changes even for the past 10,000 years involves the combination and comparison of data from many different scientific fields of study. There are – as we shall see – genuine historical records, mainly from China, covering a surprising span of this period. But the overall global picture depends on studies of the movement of glaciers, revealed by the scars they leave in the rocks and the debris they dump as they melt; on measurements of subtle changes in the composition of the ice being laid down, year by year, in layers on the Greenland glacier; on counting the pollen grains in sediments from old lake beds to find what species of plants flourished around there thousands of years ago; on counting tree rings and measuring their width to find out which years in the historical past were good growing years for trees and which were bad; and on many other techniques besides. Some of these techniques will be described in a little more detail as the story unfolds, but the story itself is concerned with the climatic changes themselves rather than with the techniques by which those changes are measured, so those details have to take a back seat. The best place to find out more about the evidence of past weather and climate fluctuations, and the history of climate changes, is in volume 2 of Hubert Lamb's epic work *Climate: Present, Past and Future*; but since this runs to 835 pages (volume 1, Professor Lamb's description of climate now, is a mere 613-page taster) I can hardly do justice to the story in one chapter here, even if I do gloss over some of the details. But the broad picture that emerges from this work can be summarized simply. There have been four distinct periods with their own characteristic climate patterns in the 10,000-odd years since the end of the most recent Ice Age, and the warmest phase of the present interglacial happened as long as 6,000 years ago.

The four climatic periods identified by Professor Lamb start with the warm epoch that followed the latest Ice Age, causing its end, and which peaked between 5,000 and 7,000 years ago. From 5,000 to 3,000 BC, sea-level rose rapidly as the ice sheets melted, the climate of the Sahara was wetter than it is today, and temperatures in Europe and North America were about 2–3° C warmer than the present-day average. This climatic optimum, as it is sometimes called (on the assumption that warmer conditions than those of today would be more pleasant), was followed by a colder epoch which corresponds very closely to the Iron Age and was at its worst between about 2,300 and 2,900 years ago. This brought not just cooling but a great increase in wetness across northern Europe from Ireland to Scandinavia and into Russia, where the great gloomy forests spread southwards as the summer temperatures fell.

After the Iron Age cold period, the next climatic landmark is provided by a warm interval, less pronounced than the postglacial optimum and therefore known as the little climatic optimum, which reached a peak in the early Middle Ages, about 800 to 1,000 years ago. By this time – AD 1000 to 1200 – we have good historical records, as well as more easily interpreted archaeological and geological evidence, which combine to show that summer temperatures

were about 1°C warmer than the present average in Europe and
North America, with vines being cultivated 3° to 5° of latitude
further north than now, and 100 to 200 metres higher above sea-
level. The last of the four main climatic periods of the interglacial is
called, appropriately, the Little Ice Age. This spell of cold, more
severe than anything else since the Ice Age proper, was at its most
extreme from about 550 to 125 years ago, but some authorities
believe that it has not yet ended, and that just as the past 10,000
years of warmth marks a temporary interval between Ice Ages, so
the twentieth century may mark, on a shorter time-scale, a temporary
respite between Little Ice Ages. From the fifteenth century to the
nineteenth, but with worst conditions in the seventeenth century, the
Little Ice Age brought an extension of the Arctic pack-ice far beyond
its boundaries during the little optimum, a general shift of climatic
belts towards the equator in the Northern Hemisphere, and a variety
of troubles for humanity. These can best be seen in perspective by
looking back, not over the whole of the present interglacial, but over
the 5,000 years or so of the historical record, from the time of Egypt
and Ancient Greece to the global society of the twentieth century.

The 5,000-year link

The link, across thousands of years of history, is provided by
changes in the circumpolar vortex and the track of rain-bearing
westerly winds, not this time across the Sahel region of Africa but
across the Mediterranean and Greece. Reid Bryson, perhaps the
most outspoken of those modern climatologists who fear an imminent return of the Little Ice Age, has taken up a suggestion made
by the classical scholar Rhys Carpenter, who argued that the decline
of the great Mycenaean civilization, which flourished in the Aegean
for hundreds of years in the second millennium before Christ, was a
result not of invasion and war but simply of drought – a change in the
weather patterns that occurred abruptly around 1200 BC. In his
book *Climates of Hunger* Bryson describes the way climatologists at the
University of Wisconsin–Madison used a computer to simulate
rainfall patterns over Greece today and to find out how these
changed when the paths of the storm tracks varied. They found that
a northward shift of the storm track would produce exactly the
drought pattern that Rhys Carpenter needed to explain the collapse

of the Mycenaean civilization, and that exactly the same shift in the circumpolar vortex, around 1200 BC, could account for droughts and famines in the Hittites' empire on the Anatolian plateau in Asia Minor. The Hittites were forced to move to what is now northern Syria – and those same computer simulations show that the changes in circumpolar patterns that brought drought to Mycenae almost certainly brought an increase in rainfall of as much as 40 per cent to this region of north-western Syria and central Turkey.

Of course, where we are dealing with such ancient history the case can never be proved, one way or the other. But it is clear that the pattern of major migrations and declining civilizations in the eastern Mediterranean just after 1200 BC exactly fits the pattern of rainfall changes associated with a slight contraction of the circumpolar vortex (Figure 2.2). Other indications from classical times provide less ambiguous evidence of the changing climate: Lamb, for example, mentions Roman goldmines high up in the Austrian Alps which are only now being discovered as the ice retreats after the recent Little Ice Age. Obviously the Roman miners were not troubled by glaciers, or they could never have worked the mines – and in the last century BC Roman agricultural writers described how cultivation of the olive and vine was spreading further north in Italy, into regions where in the previous century winters had been too cold for transplanted stock from the south to survive.

Coming closer to home, in both time and space, the next warm interval, the little optimum, provides a 'goldmine' of a different kind for climatologists, with an almost embarrassing wealth of information about the way climatic changes affected the people of the period around AD 1000. The warmth in Europe around AD 1000 to 1300 was accompanied by mild, wet winters which swelled the rivers, just as happened during a lesser mild spell in the first century AD. Providing a link between Roman times and the Middle Ages, these climatic shifts explain the otherwise baffling existence of Roman bridges across Arabian gullies and wadis that are now dry, and the remarkable five-arched bridge, the Ponte del Ammiraglio, built at Palermo, Sicily, in AD 1113 to span a river which was then navigable but is scarcely a trickle today. But for anyone who lives in the regions bordering on the North Atlantic, the sagas of the Norse seafarers provide one of the most dramatic indications of how the climate of the time differed from that of the present day. The story also provides

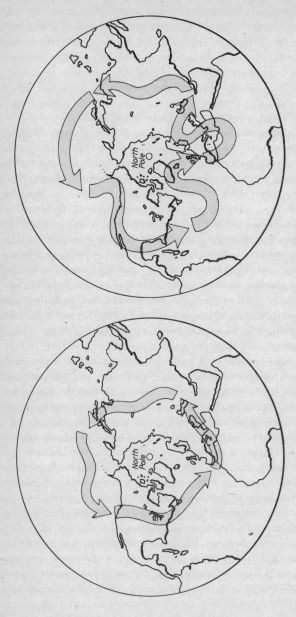

Figure 2.2. Like a snake chasing its own tail, the pattern of westerly winds wriggles around the globe, following the track of the high-altitude jet stream. Strong circulation corresponds to a tight circumpolar vortex; weak circulation corresponds to a more erratically wriggling jet stream. Such patterns affect the climate of the whole Northern Hemisphere.

an opportunity to fill in some of the details of one of the most important techniques for unravelling past climatic changes, the study of the oxygen isotopes locked up as ice in the Greenland ice-cap.

Viking triumphs

The record in the ice provides a thermometer to measure year-to-year variations in temperature over hundreds of years, because oxygen comes in two common varieties, called isotopes. By far the more common of the two is oxygen-16, which weighs 16 units on the atomic scale of measurement; its rarer and heavier counterpart is oxygen-18, which weighs two units more but is chemically identical to oxygen-16. Both types of oxygen atom combine with hydrogen to make water molecules (H_2O), so that water in the sea also comes in two varieties, one weighing two units more than the other does. The water that gets into the air to make clouds and fall as rain or snow must also contain both types of molecule, but more energy is needed to evaporate heavier molecules than to evaporate lighter ones. The result is that the proportion of the two varieties which evaporates each year, and therefore the proportion of the two varieties laid down on the Greenland ice-cap when snow falls, depends on the average temperature that year. This would be no more than a fascinating oddity of nature if it were not for the fact that the snow which falls on the ice-cap each year forms its own distinctive layer, squeezed into ice as more snow falls on top, so that the age of ice samples from deep within the ice-cap can be read by counting layers downwards from the top, in much the same way that the age of a tree can be read by counting the layers of wood laid down in its annual growth rings. The result is that by drilling a core of ice from the ice-cap, then counting the layers and analysing the isotopic composition of the ice from different layers, it is possible to determine the average temperature of any year during the time the ice was being laid down.

Of course, the technique is laborious and involves painstaking chemical analysis. But the Dane Willi Dansgaard and his colleagues have been able to analyse in this way a core of ice 404 metres long, drilled from the Greenland ice-cap, which covers a span of 1,420 years of history. This is long enough to provide a year-by-year description of temperature changes during the little optimum and the Little Ice Age – and Dansgaard's team has compared this with a detailed account of the development and demise of the Norse colonies in Greenland and Iceland, culled from the Landnam and Greenlander sagas and handed down for more than a thousand years.

Leaving aside for the moment the exact cause of these climatic changes, their impact on Norse society is clear. The first recorded attempt to settle in Iceland was made by a farmer called Floke Vilgerdson in AD 865. The attempt failed because, although he could not know it, Vilgerdson had picked the tail-end of a series of minor cold fluctuations and he lost all his cattle in a severe winter. Returning to Norway, he told how he had encountered 'a fjord filled up by sea-ice ... therefore he called the country Iceland' (Landnam Saga). Only nine years later, in 874, successful settlers reached Iceland and flourished in the rapid climatic warming of the little optimum, which shows up clearly in the isotope record (Figure 2.3). A century later, in 985, the colony was so secure that it was able to provide a springboard for further colonizing adventures when the pioneer Eirik the Red founded a daughter colony in Greenland. The Greenlander Saga tells us that Eirik chose the name as a deliberate confidence trick to lure followers to that icy and inhospitable island; but the saga probably does him an injustice, since the evidence of the isotopes from the ice-cap is that Eirik reached Greenland towards the end of a warm period longer than any that has occurred there since, when the coastal fringe, at least, may genuinely have appeared green and bountiful by the harsh standards the Viking adventurers were used to. Although neither Greenland nor Iceland is a particularly

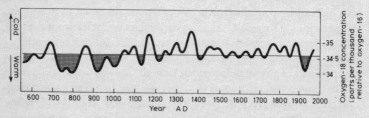

Figure 2.3. Measurements of oxygen isotope ratios in ice cores drilled from the Greenland ice-cap provide a sensitive guide to temperature variations of the North Atlantic region in historical times. These correlate strikingly with historical events such as the colonization of Greenland by the Norse, and the subsequent failure of the colony as the climate deteriorated. (Based on data from Willi Dansgaard's team, cited in the text.)

welcoming place, it would certainly make more sense today if their names were swapped; it seems they got their misleading monikers largely because of the vagaries of climatic change at the time of their discovery and settlement!

One of the first merchants plying the new route from Iceland to Greenland, Bjarni Herjolfsson, was blown so far off course that he reached America instead, but did not pause for a landing. That was left for Leif Eiriksson, the son of Eirik the Red, in the last decade of the tenth century. The tale of Vinland, almost unimaginably fertile and pleasant compared with Greenland, is now a familiar one, and it seems certain that some Norse settlements were established on the North American mainland. But by 1200 the climatic pattern had begun to change and by the fourteenth century the Greenland colony had collapsed as a result of the onset of the Little Ice Age. Having no contact with their homelands, any surviving Norse in North America either died out or were killed or absorbed by native American tribes. Even the Iceland colony, much closer to Europe, survived only by the skin of its teeth through the worst ravages of the Little Ice Age, sustained only by its value as a base for the whaling ships of the time. Before about 1200 oats and barley were grown in Iceland, but at the same time that the sagas first tell of difficulties with the colonies to the west, and increasing ice in the East Greenland Current, oats were being given up and the amount of barley harvested each year declined by 50 per cent.

North American patterns

The interlinked pattern of climatic changes also affected mainland North America, and in ways which may be of great significance today. Archaeological evidence from the western Great Plains region, the upper Mississippi valley and the dry south-western region of what is now the United States shows that there had been a wetter climate than today for perhaps as long as 500 years before the climatic shift which occurred in AD 1200. While the Norse benefited from a retreat of the sea-ice and milder conditions in the North Atlantic, native American tribes were moving northwards and westwards into present-day Wisconsin before 800, with agricultural settlements spreading up the valleys of eastern Minnesota. The Mill Creek people were one of many farming groups established on the plains on the eastern side of the Rockies by around 900, and by 1200 this was a relatively rich society, the people living in a region of tall-grass prairie with wooded valleys. They grew corn and hunted deer, living in settled communities – a strong contrast with the stereo-typical nomadic way of 'Red Indian' life portrayed in the movies.

All this changed, however, in the space of only a couple of decades at the beginning of the thirteenth century. Bryson, who lives and works in Wisconsin, has made a special study of the climatic reasons for the dramatic change in the way of life of the inhabitants of the Mill Creek region some 800 years ago. Once again, the story is one of an expansion of the circumpolar vortex, pushing the rain-bearing westerlies away southwards. As the rains disappeared the tall grass died out and was replaced by short grass, while over a longer period of time even the trees died and the plains became more open. The deer left the region, the farmers' crops failed and the settled villages of wealthy agriculturalists were replaced by a transient hunting society dependent on the bison as a reliable source of food, as well as of hides and bone. There is far more than parochial concern in Bryson's interest in the fate of the Mill Creek people, for the archaeo-logical evidence shows that they were just one set of the victims of a drought which lasted for two centuries across what is today the key grain-producing region of North America, the bread-basket of the world. The pattern of rainfall changes associated with expansion of the circumpolar vortex is seen in the archaeological record of the time across America, and extends to southern Illinois, where around AD 1000 there existed what Bryson describes as 'one of the major population centres of the world', a civilization that grew corn along the fertile Mississippi and had the resources to build great earth-works reminiscent of the pyramids of Egypt.

Although these mounds were not built of stone, their existence tells us that, like ancient Egypt, this native American civilization was rich enough to be able to feed a labour force engaged in work of no immediate practical value, and that there was a power structure with a central authority strong enough to ensure that the work was carried out. The greatest surviving structure from the time, Monks Mound, stands 100 feet above the surrounding plain and measures 1,000 feet by 700 feet; the region around Monks Mound, now called Cahokia, was the centre of a thriving civilization from AD 600 to 1200 – a rather longer span, as Bryson reminds us, than from the time of Columbus to the present day. The Cahokia people too were victims of the Little Ice Age – although, as in the case of the Mill Creek people, a more appropriate name from their point of view would be Great Drought Age.

The Thames freezes

We shall return again to the grasslands of North America, for this is the region that produces virtually all the food traded on world markets or offered as aid to the needy today. The reason why the Little Ice Age got its name, however, is best seen on the other side of the Atlantic, where one of the most reliable climatic indicators in the historical record is provided by the freezing of the River Thames. In the first thousand years after Christ the great river froze only about once per century at the site of London. In 1209 the famous old London Bridge was built, and from then on a combination of the bridge and changing climate altered this pattern dramatically.

The sturdy construction of the bridge played its part, since there was more bridge than there were gaps for the water to pass through, with the result that the water collected upstream, pouring through the arches like a waterfall when the river was high. In winter, with branches and other debris also piling up at the bridge, any ice that formed soon froze into a solid platform and spread upstream, growing on many occasions several feet thick and solid enough not just to walk on but for horses and carriages to be driven across the Thames. From the beginning of the thirteenth century onwards the records of the frozen Thames provide a clear guide to the development of the Little Ice Age.

In 1269-70 the Thames froze not just upstream from the bridge, but so far downstream that goods normally transported from the coast in boats had to be sent overland; lesser freeze-ups in 1281-2 and 1309-10 were followed by a century of not-too-severe weather, but between 1407-8 and 1564-5 the river froze no fewer than six times. Henry VIII drove on the river in either 1536-7 or 1537-8 (the records are unclear on the exact date) and Elizabeth I used to take a regular stroll on the icy Thames in the winter of 1564. In the seventeenth century the river began to be used almost as a regular winter sports arena; the first 'Frost Fair' was set up in a tented 'city' on the ice in 1607-8. Booths in the tent 'city' sold food, beer and wine; there was bowling, shooting and dancing on the ice by way of entertainment. There was another frost fair in 1620–21, and the sport of skating was introduced to England in 1662–3 when the Thames froze again. The greatest of all frost fairs was held in 1683–4, and another took place in 1688–9. Altogether, the river froze ten times in the seventeenth century and ten more times between 1708–9 and 1813–14; but it has never frozen completely since. This is partly because the old bridge is no longer there, partly because industrial pollution and waste heat pumped into the river mean that it could not have frozen even if a winter as severe as 1683–4 had struck in the late nineteenth century, but also because of the significant warming trend that ushered in the twentieth century.

A similar pattern is seen in records from across Europe, with many lakes and rivers freezing in the worst decades of the Little Ice Age, especially during the seventeenth century. Further south, as in North America, the story is one of changing rainfall rather than changing temperature. In the Middle Ages, North Africa, including the fringes of the Sahara, was wetter than today, but during the Little Ice Age it was drier. Just which climatic changes different experts pick out as significant depends on their personal interests; what matters far more is that all the changes fit into a clear pattern involving changes in the overall circulation of the Northern Hemisphere (the Southern Hemisphere changed too, but records from the south are far more sketchy). The Little Ice Age, in spite of its name, was not a time of unrelieved cold, and many summers in Europe in the second half of the seventeenth century were hot and dry, with drought a real problem. Both plague and the Great Fire of London, which broke out in 1666 and raged through the city for five days, destroying 14,000 buildings and leaving 200,000 people homeless, are associated historically with the Little Ice Age droughts.

Hubert Lamb has summed up all the evidence and concludes that in the summers of the little optimum period the main flow of the strong westerly winds in the upper atmosphere ran 3° to 5° further north than today, whereas in the coldest period from 1550 to 1700 the mainstream of the circumpolar vortex ran 5° further south than in the middle of the twentieth century. England and Europe were wetter overall as a result, although there were drought years, and the Great Plains of North America, as we have seen, dried out in the rain shadow of the Rocky Mountains. The trend of the Little Ice Age in Europe was towards wetter summers and more severe winters, a pattern as familiar to Europeans in the early 1980s as drought is to North Americans today. But there was also, in Lamb's words (p. 465), 'enhanced variability of temperature from spell to spell, from year to year and on somewhat longer time scales. *This is characteristic of regimes with frequent blocking* ... [my italics]. Despite some record cold spells, periods of great heat in England were also recorded.' The overall picture, says Lamb, shows the Little Ice Age to have been a time of 'atmospheric circulation patterns associated with an expanded polar cap, expanded circumpolar vortex, frequent blocking...'. It is no wonder that Bryson and his colleagues are alarmed by recent developments, which have included an expanded circumpolar vortex (bringing droughts to the Sahel) and frequent blocking, bringing among other things the hottest summer in England since records began ('periods of great heat'), severe winters in both Europe and North America ('record cold spells') and a return of drought conditions to the Great Plains. These developments echo grimly the climatic picture of the Little Ice Age.

Severe weather in Europe left its mark on society, a mark which persists to the present day. It was the ravages of the Little Ice Age that caused James VI of Scotland, by then also James I of England, to establish a settlement of his Scottish subjects in Ulster, in the north of Ireland, an act that has repercussions still in the troubles of that area. Whole villages in England, especially in East Anglia, were depopulated and ceased to exist, not because of the Black Death, as historians used to think, but because of the crop failures and starvation caused by climatic changes. Modern scrutiny of the records shows time and again that the local populations declined and suffered starvation first, and that the plague then took its toll of weakened survivors, rather than being the main cause of the catastrophe. As a result this region still has empty and underpopulated stretches today.

American droughts

Since the early nineteenth century conditions have changed again. The circumpolar vortex seems to have withdrawn a little, with all that that implies. The Little Ice Age looks like a period of cold and 'unusual' climatic extremes to us because conditions in our lifetimes have been different. But there is also direct evidence of the changing trend, a trend which Bryson describes as 'persistent and obvious'. Temperatures in central England, for example, rose for a hundred years from 1850, and the period 1925–54 was 1½° C warmer than the previous 25 years; the length of the growing season at Oxford (essentially the interval between late spring and early autumn frosts) was two to three weeks longer in the 1940s than in the nineteenth century; the average temperature at Copenhagen in the 1950s was 1½° C higher than in the 1850s; glaciers in Europe retreated; and mean temperatures in Iceland also rose by about 1½° C. All of this, says Bryson, makes it clear 'that the early and middle years of our own century – the time on which we base our ideas of normal climate – have been quite different from the 300 years that came before. We have had one of the least normal climate patterns of the last 1,000 years, *not* the only pattern ... the illusion that climate is stable and unchanging is finally becoming untenable' (pp. 90–91, *Climates of Hunger*).

This makes it doubly frustrating that accurate modern records for the most crucial agricultural region in the world today, the Great Plains, only began to be kept in the second half of the nineteenth century. But we can get some idea of the real normal climate of that all-important region, from the perspective of the past thousand years or so of climate, by studying the history of western exploration, beginning with the expedition of Major Stephen H. Long in 1819–20. The most telling feature of Long's report on the high plains is the name he gave them, based on his first impressions – the Great American Desert.

The Long expedition was followed by others, but the region remained largely unsettled until the 1850s, when Lieutenant G. K. Warren reported back to Washington that the western limit of agricultural land could be identified with the 97th meridian, which runs roughly through Wichita, Kansas, and Lincoln, Nebraska. Warren described the lands west of 97° as a 'desert space' separating the eastern agricultural zone from the fertile lands of the Pacific Coast, and Major John Wesley Powell, the first Director of the US Geological Survey, endorsed this view in his 'Report on the Lands of the Arid Region of the United States', issued in 1875, which stated that there was insufficient rainfall to support crops without irrigation west of the 100th meridian.

So just over 100 years ago, and just after the Civil War, scientific opinion unanimously confirmed the reality of the Great American Desert. Pioneers pushing westwards with the railroad took little notice of scientific opinion, however, and in the optimistic, expansive mood of the nation following the Civil War, the catch-phrase on many people's lips was the 'manifest destiny' of the nation to expand westwards. The pioneers were lucky – at first. In the 1880s, roughly coinciding with the establishment of a network of meteorological recording stations, the rainfall of the high plains region, though never reliable, increased slightly, giving rise to the bizarre theory that 'rain follows the plough', and that by opening up the region to agriculture the pioneers were in some way stimulating the increased rainfall. But even by 1889 the writing was on the wall as dry conditions returned to the plains, and severe drought in the early 1890s led to the abandonment of many homesteads and the creation of ghost towns. Eastern farming techniques had, indeed, proved impracticable on the high plains.

After the droughts eased, a new wave of homesteaders moved in again, and through the early part of the twentieth century the plains were farmed with some success, not by traditional eastern methods but by cultivating huge tracts of grassland and making increased use of mechanization. It looked as if technology could overcome the arid conditions – until the 1930s and the era of the great dust-bowl. Plains that used to be protected by a tough mat of grass had long since come under the plough and lay unprotected at the mercy of winds which blew away the fertile topsoil for hundreds of miles. The plains have never really recovered from that drought, in spite of efforts to replace the grass cover and plant trees as windbreaks, efforts handicapped by further droughts in the 1950s and 1970s. We shall hear more of this roughly 20-year drought cycle later, together with increasing scientific evidence that the cycle is not over, but due to bring renewed drought conditions in the Great American Desert in the early 1990s, less than ten years from now.

Detailed records prior to 1872, when the US Weather Bureau (now the US Weather Service) was founded, are hard to come by, but all the evidence is that during the twentieth century a region that had been arid and unsuitable for agriculture improved, climatically, just enough to make farming possible in the good years, but not during the driest years of the rainfall cycle. The same pattern, of a slight climatic amelioration in North America in the twentieth century, emerges from the limited temperature data available, which have been studied by meteorologists at the University of Wisconsin–Madison. Some temperature measurements are available, for example, for Fort Winnebago, near Portage, Wisconsin, for the period 1829–42. When these are compared with temperature records from Portage covering the period from 1931 to 1948, they show that in every month of the year except March the nineteenth-century period was colder, with the effect most pronounced in August (5·1° C colder), September (6·9° C colder), October (4·5° C colder) and November (5° C colder). In addition, wind observations from Fort Winnebago show that there was a much more frequent incidence of winds from the north-west and east in the nineteenth century. All these figures exactly fit the picture of an expanded circumpolar vortex in the nineteenth century, the aftermath of the Little Ice Age. The crucial question now is: has the amelioration of the twentieth century, produced by a contraction of the circumpolar vortex, really spelled the end of the Little Ice Age? Or have recent decades simply provided a temporary respite, with the prospect of a renewed expansion of the circumpolar vortex, a return to nineteenth-century weather or even to a full Little Ice Age, due any time now?

The cooling now

There is no doubt that the early twentieth-century warming was followed by a cooling which set in across the Northern Hemisphere before 1950. Climatologists debate whether this cooling trend is continuing, or whether it may already be showing a new upturn – that debate, and the possibility that any natural cooling trend may be offset by mankind's influence on the climate, is what this book is all about. Hubert Lamb, however, describes the cooling since the early 1940s and up to at least 1975 as 'the longest-continued downward trend of this item since 1700' (p. 529), with the greatest cooling in the Arctic. By 1968–72, the five-year average of temperature in the North Atlantic measured by nine ocean weather-ships stationed between 33° and 66° N was 0·56° C lower than the equivalent mean temperature for the period 1951–5; these and other measurements, says Lamb, provide evidence of 'weakened global atmospheric circulation in the last decade or two' (p. 535). Writing just before some of the recent extreme climatic events that have since hit the headlines, he was still able to say that 'the years 1962–75 have seen many temperature records in the northern hemisphere middle latitudes of both high and low extremes which could only be matched, if at all, about 200 to 220 years earlier'.

This comment is intriguing for many reasons. First, as I hope the evidence in this chapter and the last makes clear, extremes of *both* kinds are more common when the Earth cools and the global vortex is weak, because then blocking highs can become established. When the world is warmer and the circulation is strong but more tightly following a circular path around the pole, conditions tend to be more even-tempered, with fewer extremes of temperatures and more predictable rains for the farmers in the middle latitudes. Secondly, although more tenuous than most of the evidence of climatic variability I have mentioned so far, there is at least a hint from long climatic records such as the isotopes in the Greenland ice core of a repeating rhythm in climate, as revealed by temperature fluctuations, that is roughly 200 years long, perhaps a little less. And thirdly, Lamb's casual – almost throw-away – remark that the period from 1962–75 marked a series of climatic extremes unprecedented for 200 years eerily sets the scene for a study carried out by John Kington of the University of East Anglia. This was reported late in 1979 and showed that the climatic pattern of the 1780s, when studied in detail, shows more similarity to the pattern of the ten years from 1968 to 1978 than to any other decade in the intervening period. Turning that description round, Kington's study tells us that the 1970s were more similar to a typical decade of the latter part of the Little Ice Age than to any decade of the subsequent climatic amelioration.

It is worth looking at this analysis in a little more detail, since it suggests so clearly that Little Ice Age conditions are indeed returning. Unfortunately, in view of the importance of US grain to the world food supply today, the evidence all comes from Europe – there were no meteorologists on the high plains in the 1780s! But since it is now clear that the workings of the atmosphere are interconnected, with the key features on a time-scale of decades or centuries being changes in the circumpolar vortex, the European evidence is good enough to sound a warning note.

Eighteenth-century weather is back

In the summer of 1979 a conference held at the University of East Anglia in Norwich, England, saw the first formal coming together of historians and climatologists from around the world to look at the puzzle of climatic changes in historical times and their influence on mankind. At that meeting, faced with evidence of the kind presented in this chapter, but in far greater detail, many of the climatologists emerged with a consensus that although the warming effect of a build-up of carbon dioxide in the atmosphere – the greenhouse effect – may soon become a real influence on climate, the immediate threat, for the next decade at least, is of a continuation of the shift back into the cooler and more erratic weather conditions of the Little Ice Age. Professor Lamb pointed out that on the available evidence it seemed that planners preparing for the likely climatic extremes of the immediate future – the next ten years, say – could profitably study the climatic records of the seventeenth and eighteenth centuries, rather than relying on the more easily accessible and detailed records of the 'normal' climate of the past hundred years or so. John Kington's study put the icing on the cake of Lamb's summary of the situation.

Kington's study is part of a continuing long-term effort by the University of East Anglia climatologists to plug a gap in our knowledge of past climates, from the time of the invention of the thermometer and barometer in the early seventeenth century to the introduction of the electric telegraph in the 1850s. Before the 1850s it was impossible for weather forecasting to be attempted in the way it is nowadays, using maps of the distribution of pressure and temperature today (synoptic weather charts) as a basis for predicting what the weather will be like tomorrow. Throughout the eighteenth century, ever-improving records of temperature and pressure were kept at sites across Europe – but in the days before the telegraph a meteorologist in London, say, could not put this together quickly enough to be useful for forecasting. Stage coach or carrier pigeon might get the information to him, and he could use the information to draw up a weather map; but by the time he did so the map would be several days out of date! So the century before the introduction of the telegraph provides rich pickings for students of historical climatology. The records needed to construct weather maps are there, but by and large the maps have never been constructed.

Kington's team aims eventually to fill this gap by reconstructing synoptic charts for the entire period, doubling the length of the record we have of detailed variations in the weather over Europe and providing a much better guide to what really is normal weather. The team chose to start their survey with the decade of the 1780s, partly because the first attempts at reconstructing historical weather patterns were made for the 1780s by the German meteorologist H. W. Brandes in 1820, and partly because the historical evidence already hinted that the decade had been marked by some interesting climatic extremes. When work on the project began in Norwich in the early 1970s the team had no idea that the world was even then in the grip of a similarly remarkable decade of climatic variability.

Although the data are available, however, the historians have no easy task in obtaining them and putting them into usable form. Detailed sources include records of the Société Royale de Médecine (SRM) in France, a network of observing stations established to investigate possible effects of weather and climate on illnesses and epidemics. In Germany, Elector Karl Theodor of the Rhineland Palatinate established in 1780 an observing network under the organization of the Societas Meteorologica Palatina (SMP), using standardized instruments, and this network eventually spread across Europe as far as Russia and over the Atlantic to Greenland and North America. In Britain and Scandinavia official observations were already being made, along with the many regular observations made on a personal basis and kept in private logs and diaries. The preserved logs of British and other naval vessels also provide day-to-day (even hour-to-hour) observations of weather conditions, and these are particularly useful as they extend coverage out to sea. Finally, accounts from farmers and traders help to fill in gaps in the detailed record.

Collecting all this data has involved the Climatic Research Unit at the University of East Anglia in searches of libraries across Europe, as well as extensive correspondence with both libraries and individuals. The result is a mass of material, often handwritten, in several different languages, in different formats, using different measuring units, and of different quality. It is hardly surprising that although the project was conceived just over ten years ago the first detailed results have only recently been published, for the first half-decade studied. The team deserves their stroke of serendipity that the 1780s survey turns out to tell us a great deal about what is happening to our weather today; although similar studies of other decades will follow more quickly now the groundwork has been done, at the time of writing the 1780s study remains a unique piece of historical reconstruction, worth special mention in its own right, and doubly so, it turns out, as a mirror of the 1970s.

So far, six complete years of daily weather charts are available,

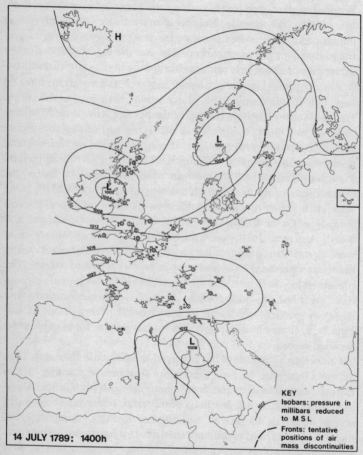

14 JULY 1789: 1400h

KEY
Isobars: pressure in millibars reduced to M S L

Fronts: tentative positions of air mass discontinuities

Figure 2.4. 'Well suited to outdoor activity'. John Kington's reconstruction of the weather chart for the day of the storming of the Bastille.

covering the years 1781–6, together with some short sequences of specially prepared charts providing case studies of interesting climatic or historical events, such as the weather on the day of the storming of the Bastille (14 July 1789), which Kington describes as 'fine and dry, well suited to outdoor activity' (Figure 2.4), and a sequence from around the time of the sea battle on 23 September 1779 in which John Paul Jones's *Bonhomme Richard* clashed with the British frigate *Serapis* off Flamborough Head (Figure 2.5). With more than 2,000 synoptic charts available for statistical analysis and interpretation, the material has to be converted into a more digestible form, and the team does this by classifying the charts by the weather types they represent, in accordance with a now standard scheme devised by Professor Lamb (and described in Volume 1 of his epic work on climate).

In a nutshell, Lamb's classification scheme relates the direction of prevailing winds at any time to the occurrence of high (anticyclonic) or low (cyclonic) pressure, and relates these to the changing patterns of the circumpolar vortex. For Britain, more 'westerly' weather implies a regular movement of depressions and their burden of rain in from the Atlantic, and corresponds to less severe winters. This matches up with the strong circulation pattern and a contracted circumpolar vortex. With less westerly weather Britain experiences longer-lasting extremes of 'continental' weather, including both summer droughts and severe winters, roughly corresponding to the weak circulation pattern with an expanded, erratically zig-zagging circumpolar vortex and more frequent blocking states.

As the figures of Table 1 show, the pattern of daily weather types for 1781–5 bears very little resemblance to the pattern from 1861 to 1969, the period over which 'normal weather' is defined. The average value of 66 days per year of westerly weather over the earlier five-year period is remarkably low compared with the average of 93 days per year for the century before 1970, an average which rose to 100 days per year for the period 1900–1950. Since the 1950s this index of weather variability has shifted again, averaging 80 days per year in the 1960s and just 66 days per year – very close to the 1780s pattern – in the four years up to 1971. Picking just one example from the mass of information, Kington points out that the longest period of drought in England and Wales prior to the record-breaking events of 1975–6, mentioned in Chapter 1, was from August 1784 to July

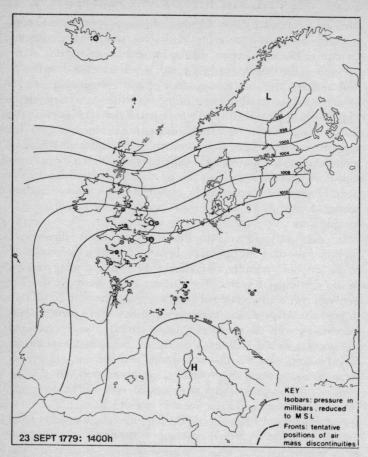

KEY
Isobars: pressure in millibars, reduced to M S L

Fronts: tentative positions of air mass discontinuities

23 SEPT 1779: 1400h

Figure 2.5. In an attempt to locate the site of the sunken Bonhomme Richard, *a US team commissioned a special study by John Kington of the weather in the North Sea in September 1779. On Thursday 23 September after morning cloud and rain the weather cleared by midday to become warm and pleasant. There was decreasing wind, a gentle breeze (south-west, force 3), becoming calm in the evening with smooth seas. So far, the attempt to reconstruct the course and ultimate resting place of John Paul Jones's ship has been unsuccessful, but the search continues with the aid of charts like this, and may yet achieve its goal.*

Table 1: British Isles daily weather types, yearly and average frequencies, 1781–5 and long-period averages, 1861–1969

	W	NW	N	E	S	A	C	U	W	NW	N	E	S	A	C	U
	Number of days								*per cent*							
1781	$74\frac{3}{4}$	28	$19\frac{1}{4}$	$37\frac{2}{3}$	$29\frac{5}{6}$	91	$77\frac{1}{2}$	7	20·5	7·7	5·3	10·3	8·2	24·9	21·2	1·9
1782	$75\frac{5}{6}$	27	$31\frac{1}{3}$	43	24	$63\frac{1}{6}$	$90\frac{2}{3}$	10	20·8	7·4	8·6	11·8	6·6	17·3	24·8	2·7
1783	$75\frac{1}{2}$	13	$21\frac{1}{2}$	$24\frac{1}{2}$	$35\frac{1}{2}$	$98\frac{5}{6}$	$84\frac{1}{4}$	12	20·7	3·6	5·9	6·7	9·7	27·1	23·1	3·3
1784	$59\frac{2}{3}$	$27\frac{1}{2}$	$46\frac{1}{6}$	$42\frac{1}{2}$	$20\frac{1}{3}$	$106\frac{1}{2}$	$56\frac{1}{2}$	7	16·3	7·5	12·6	11·6	5·5	29·1	15·4	1·9
1785	45	$32\frac{1}{2}$	$35\frac{1}{2}$	$30\frac{1}{6}$	$30\frac{1}{6}$	$97\frac{2}{3}$	80	14	12·3	8·9	9·7	8·3	8·3	26·8	21·9	3·8
Average (1781–1785)	66·2	25·6	30·7	35·5	28·0	91·4	77·8	10	18·1	7·0	8·4	9·7	7·7	25·0	21·3	2·7
Average (1861–1969)	93	18	27	28	31	91	64	13	25·5	4·9	7·4	7·7	8·5	24·9	17·5	3·6

W Westerly type
NW North Westerly type
N Northerly
E Easterly type
S Southerly type
A Anticyclonic type
C Cyclonic type
U Unclassifiable

1785. 'The main contribution of this project to present-day affairs', he says, 'is that it is helping to provide a fuller picture of climatic behaviour within the past 200 years. This information can assist in national and economic planning [and] in being better prepared to anticipate future outstanding climatic events, at a time when there is a general decline of the westerly weather type' – that is, at a time, today, when there is a shift away from the strong circulation pattern that dominated the mid twentieth century, and back towards the Little Ice Age conditions that have dominated the past millennium.

The historical perspective of the past thousand years thus answers the puzzle of why we live in such interesting times where climate and weather are concerned. On a time-scale of centuries, the early to middle part of the twentieth century, the period over which 'normal' weather patterns were defined, was in fact far from normal and represented the most unusual fifty-year period in the climatic record of the past thousand years. The downturn in temperatures, with its related climatic changes, since the 1940s shows clearly that this warming was a temporary deviation from the Little Ice Age pattern, not a breaking of the mould; and the weather extremes of the 1970s and early 1980s are a sign that truly normal weather – eighteenth-century weather – is back. Little Ice Age conditions, drought on the high plains and in the Sahel, shorter growing seasons in temperate latitudes around the globe, and all that these imply, are the natural patterns we can expect in the decades ahead – unless and until human activities alter the natural balance of climate and weather.

But why does the climate change? Without some understanding of the way the global weather machine works, and how and why it changes gear, forecasting cannot be other than imprecise. And only by knowing how the natural checks and balances operate can we really understand how mankind's activities may indeed be un-balancing the weather machine. Fortunately, the meteorological sciences have advanced to the point of being able to explain these natural fluctuations just in time for the knowledge to be used as a guide to anthropogenic influences on climate. The story begins with the changing pattern of atmospheric circulation which structures the climate zones of our planet and brings us the cyclic rhythm of the seasons.

3
The Weather Machine

The general circulation pattern of the atmosphere – the weather machine of planet Earth – is set in motion by inequalities in the amount of heat being received from the Sun at different latitudes. This pattern is then complicated by the different ways oceans and continents respond to the incoming solar heat, and by the rotation of the Earth, which sets the whole great convection pattern swirling in a more complex fashion. Climatic changes, whether on a scale from Ice Age to interglacial and back again, or from adequate rainfall into a time of drought on the great plains of North America or in the Sahel, depend on minor modifications to the smooth running of the weather machine – not even the equivalent of a spanner being thrown into the works, but more the tiny interruption which might be caused, to extend the analogy, by a piece of grit in one of the cogwheels. And the only way to understand these minor fluctuations is to have at least a broad outline of the way the whole machine works.

Solar energy is the key to an understanding of weather and climate. The Earth receives heat from the Sun, and it radiates heat out into space. The two processes are exactly in balance in the long term, since otherwise the Earth would either be steadily heating up, if it absorbed more energy than it radiated, or steadily cooling down, if it radiated away more heat than it received from the Sun. (A little heat does escape from the interior of the Earth, especially in volcanic regions such as Iceland. Over the whole planet, however, this has a negligible effect compared with heat from the Sun.)

The terms 'energy' and 'heat' are interchangeable here; more accurately, we should talk about electromagnetic energy from the Sun, which radiates mainly in the part of the electromagnetic spectrum we see as visible light. This is no coincidence – our eyes have evolved and adapted over hundreds of millions of years to respond most sensitively to the kind of radiation the Sun supplies. But solar radiation does extend a little bit either side of the spectrum of visible light, at the short-wavelength end beyond the blue and violet part of the spectrum into the ultra-violet, and at the long-wave end beyond the red part of the spectrum into the infra-red. Infra-red radiation – the infra-red region of the electromagnetic spectrum – corresponds to the heat energy emitted from a cooler surface than that of the Sun, and most of the Earth's radiant energy is in this form. The surface temperature of the Sun is close to 6,000° C, while the average surface temperature of the Earth is about 15° C. This difference alone determines the wavelength at which most energy from each of the two is radiated, so that the heat balance of the Earth is struck between incoming solar energy at the wavelengths around visible light and outgoing terrestrial energy at infra-red wavelengths. The difference is crucial to the nature of the Earth, as we shall see; in terms of driving the weather machine, however, the first important effect is that much more heat arrives at the equator on a particular area of ground (or sea) surface than at high latitudes. The general circulation of the atmosphere is a result of the natural processes which tend to even this difference out as far as possible.

But the averaging out is not done with radiant energy alone. Energy can move from one place to another in three different ways – across empty space from the Sun to the Earth, it travels as radiant energy, just the same as the warmth which reaches us from glowing coals or the logs of an open camp fire. Through a solid material, such as the metal of a poker used to stir the embers of a fire, heat travels by conduction, as the atoms in the solid jostle one another into more vigorous motion. Neither of these processes, however, provides the driving force for the atmospheric weather machine. The atmosphere is too diffuse a collection of molecules for conduction to play a big part in moving heat around the globe, and it is also transparent to solar radiation. The incoming radiation warms the surface of the Earth and heat passes from the warmed surface to the lower layers of the atmosphere, partly by conduction but mainly through infra-red radiation, which is absorbed by molecules such as water and carbon dioxide in the atmosphere. Once the lower layers of air are warmed, the third process of energy transfer, convection, takes over.

Convection at work

Convection is best summed up in the old adage, 'hot air rises'. At its simplest, the general circulation of the atmosphere is a result of hot air rising in the tropics and being displaced north and south of the equator as more hot air rises underneath. The displaced air cools, because it radiates heat away as infra-red energy into space and sinks at higher latitudes, where it gives up more heat to the surface of the Earth in regions which do not get the benefit of tropical sunshine. But this simple picture disguises all the interesting features of the weather machine which make our weather and climate so changeable.

First, there is a difference between the amount of heat being radiated by the Sun and arriving at the top of the atmosphere (usually called the 'solar constant', even though some astronomers believe that the heat output of the Sun itself does vary slightly) and the amount of heat getting through the atmosphere to each square metre of the ground, which is called the 'insolation'. Although the atmosphere itself is almost transparent to solar energy, clouds in the atmosphere can reflect away a great deal of incoming sunlight, while tiny dust particles scatter the incoming sunlight in all directions. This scattering makes the sky blue, because the shorter wavelengths, corresponding to blue light, are more easily scattered and bounced around to appear from all parts of the sky. Red wavelengths are less easily scattered, so they penetrate directly through even a dusty layer of the atmosphere to the ground – which is why sunsets are red. The red of a sunset is the left-over colour after the shorter wavelengths have all been scattered out.

Once the energy reaches the ground it may all be absorbed if it falls on a dark surface such as a dense tropical forest, or it may be almost entirely reflected back out into space if it falls on a shiny surface cover such as the Antarctic ice-cap. And because the amount of insolation (heat per unit area) depends on the angle the Sun is above the horizon, it varies with time of day, with latitude, and with the seasons. When the Sun is low in the sky, its incoming energy is spread thinly over a wide surface area; when it is high in the sky, the energy is concentrated and the insolation reaches peak values. It is because the Sun is always high in the sky at noon in the tropics that the equatorial regions absorb more heat than the high latitudes and set the circulation of the weather machine in motion. The high latitudes get plenty of insolation in summer, when their hemisphere

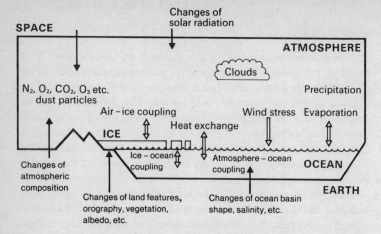

Figure 3.1. This schematic illustration, prepared by the US National Academy of Sciences, indicates the complexity of the interactions between land, sea, air and ice which affect the workings of the weather machine.

is tilted towards the Sun, but over the polar regions this is more than compensated for by the reflectivity of the snow and ice fields that are the legacy of many long, cold winters with scarcely any sunshine at all. Even when the Sun is high in the sky it cannot warm the polar regions dramatically because so much of its heat is reflected straight back into space.

The circulation patterns that define weather and climate occur almost entirely in the lowest layer of the atmosphere, the troposphere, which extends from the surface up to about 15 km at the equator and to about half that altitude over the poles. This is the thinnest layer of the atmosphere in the sense of its height (Figure 3.2), but the thickest in terms of the density of air within it, and the troposphere contains 80 per cent of the total mass of the atmosphere. Because of the way sunlight heats the Earth's surface, which then heats the air immediately above it, the troposphere is warmest at the ground (or sea) and cools off progressively with increasing altitude. The layer of atmosphere immediately above the troposphere is the stratosphere, and this layer absorbs energy directly from the incoming sunlight, because of photochemical reactions involving ozone, the triple-atom version of oxygen. As a result the stratosphere is warmer than the top of the troposphere, and the old adage about

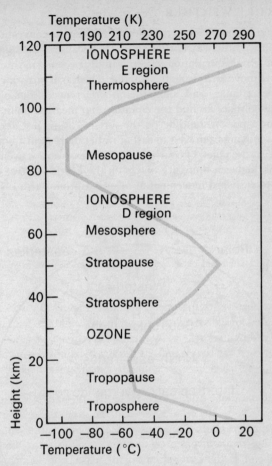

Figure 3.2. The layered structure of the atmosphere of our planet is clearly seen in terms of the way temperature changes with height. The lowest layer, the troposphere, is the region in which weather systems circulate.

hot air rising only applies if the air above is less hot. So for practical purposes convection stops at the boundary between the troposphere and stratosphere, the tropopause.

If the Earth did not rotate and had a smooth, billiard-ball surface, then this circulation would be simple indeed. But in the real world the 'ideal' pattern of warm air rising, moving away from the equator, sinking and flowing back to the equator to close the convection loop extends only a little way north and south. Because of the spin of the Earth, the simple circulation pattern rapidly becomes more complicated as we move away from the equator. The Earth turns once on its axis every 24 hours, and, since the circumference of our planet is about 25,000 miles around the equator, that means a point on the surface, or a person standing on the surface, or air just above the surface, is moving from west to east at rather more than 1,000 miles per hour. At the poles, of course, there is no eastward movement at all, and in between there are places on the Earth's surface corresponding to eastward movement at all speeds from 0 to 1,000 miles

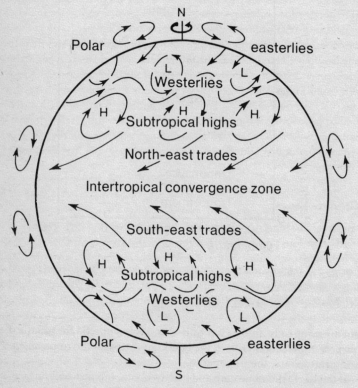

Figure 3.3. The global atmospheric circulation set in motion by the heat of the Sun and the rotation of the Earth produces latitudinal bands dominated either by high or low pressure or by the trade winds. These are the basic climatic zones of our planet.

per hour. When the tropical air which is following the convection pattern returns to the Earth's surface some way north or south of the equator, it carries with it almost its original eastward speed. But the surface of the Earth in the region outside the tropics where the descending tropical air settles is moving eastwards at a lesser speed. The result is that the descending air blows from west to east as a prevailing wind in the middle latitudes. The winds blowing back towards the equator are moving from a region of slower eastward movement into the region of fastest eastward movement, and they are also slowed by friction at the earth's surface. So the prevailing winds that blow towards the Intertropical Convergence Zone have an easterly slant, blowing in from the north-east and south-east, not due north and south across the equator (Figure 3.3).

In the Northern Hemisphere, the prevailing mid-latitude westerlies dominate the weather patterns of Europe and North America. In the Southern Hemisphere, the equivalent winds sweep around the globe over a region almost entirely free from land; unobstructed by any great mountain ranges, they become the 'roaring forties' of the days of sail. Beyond the region of descending, westerly air in each hemisphere some of the original tropical air continues its high-level journey polewards, eventually descending over the polar regions and cooling to produce bitter easterly winds that move out from the polar regions in winter.

Climate zones

All these circulation patterns can be related to the distribution of high- and low-pressure systems around the world, and to regions of high and low rainfall. Some of the Sun's heat – in fact a great deal of it – goes not just into warming the surface of the Earth in the tropics, but into evaporating water. So the hot air rising in the tropics is not just hot but moist as well, and one result of the cooling of this rising air is that the moisture goes back into the form of water droplets, giving up heat as it does so and making clouds from which the tropical rains fall. Roughly speaking, rising air is associated with rain, and that is why the tropics are wet and covered with lush vegetation. Conversely, where the airflow is mainly downwards, moisture is taken up by the airstream as it is warmed and compressed. The downward flow causes a pile-up of air at the surface, increasing the atmospheric pressure; so the overall effect is that regions of descending air are both dry and dominated by high-pressure systems. The deserts of the US south-west, the Sahara and other arid regions around the world are typical products of this feature of the atmospheric circulation; and, although we are not used to thinking of them as deserts, the dry, cold regions over the poles are produced in exactly the same way. The region dominated by the westerlies lies between these two regions of dry, sinking air. So the observed climate zones of the world fit in well with the basic understanding meteorologists have of the workings of the weather machine.

From about 5° south of the equator to 10° north (the difference is caused by the imbalance of the distribution of land over the globe) the weather is hot and wet, equatorial weather with daytime temperatures much the same all the year round, a steamy 27° C with high humidity and rainfall levels reaching more than 200 cm in a year in many places. With heat and wet combining to produce luxuriant vegetation, this is a great place for insect life, but not very pleasant for humans. Away from the equatorial zone to about 20° of latitude on either side of the equator, the weather is mainly influenced by the trade winds, the ground-level part of the returning convection circulation. Because the ICZ shifts north or south with the seasons, however, these regions sometimes come under the influence of equatorial weather systems, with a marked shift from the dry, trade-wind dominated winters to the sultry, equatorial summers. These regions are the true tropics, in weather terms, where during the dry season under clear skies temperatures can climb into the high thirties centigrade; but at night those same clear skies allow heat to be lost so rapidly that temperatures may fall below 15° C. Rainfall averages 100 to 150 cm per year, but continental interiors are much drier than coastal regions and have a stronger contrast between the seasons. Parts of the tropical zone can be very suitable for human occupation, the main problem being unpredictable droughts making agriculture difficult. The northern fringe of Australia and Central America are examples of tropical regions with good rainfall; East Africa and the Sahel are examples of regions in the same climate zone but with unreliable rainfall, sheltered by the bulk of the continent to the west from the prevailing winds.

The Sahel can also be considered part of another climate zone, the region where the tropical pattern of seasonal rainfall is so pronounced that it is called the monsoon. Polewards from the monsoon belt, there are the great deserts produced by the influence of descending air: a kind of buffer zone between regions dominated by the circulation patterns associated with the equator and those dominated by the circulation around the poles, in particular the circumpolar vortex. On the poleward side of the hot desert region, the first zone in which the westerlies dominate is described by the term 'Mediterranean climate' – logically enough, because the Mediterranean itself is the archetype of this pattern, which is found on the western sides of continents between about 30° and 40°, and is characterized by a pronounced seasonal shift of weather from hot, dry summer conditions to warm, wet winters dominated by prevailing westerly winds. Apart from the Mediterranean itself, California, south-west Australia and the tip of South Africa all have Mediterranean climates – which is why they all produce fine wines. This is, in the eyes of many people, the best climate for human habitation, with daytime temperatures perhaps a little uncomfortable in high summer, peaking about 35° C, but with pleasant evenings and cool nights, while winters are no harsher than the late spring in a northern country such as England, and rainfall, at between 40 and 80 cm, is sufficient for agriculture but tends to fall in bursts. Rainy days are really wet – but most of the time it is fine and dry.

Beyond the Mediterranean climate zone things initially become more interesting. In the temperate zone between about 40° and 60° in the north and 35° and 55° in the south the westerlies really do dominate – but 'temperate' is hardly an appropriate name, since, although the zone includes western Europe (it was western Europeans who gave it this name) and similar regions such as New Zealand, it also includes the heartlands of continents, such as Siberia and the central-northern US and Canada, where conditions are much more extreme. These far from temperate regions are, therefore, generally described as having continental climates, while the only true temperate regions are those on the western sides of continents, dominated by the arrival of successive weather systems sweeping in from the oceans further to the west. The prevailing winds off the ocean keep the temperate regions cool in summer and warmer than they would otherwise be in winter. This is an ideal climate for agriculture, but not to everyone's taste as far as human habitation is concerned.

At higher latitudes still, we come to the polar and sub-polar regions where winters are long and hard. The further towards the pole, the weaker is the compensating influence of summer and the harsher conditions become. The snowy wastes of the Arctic sea-ice and the frozen continent of Antarctica are true deserts which support no plant life, and where animal inhabitants – such as the Antarctic penguins – get their living from the sea, not from the land.

Almost an Ice Age

When we talk about the expansion or contraction of the circumpolar vortex squeezing climate zones towards the equator, or allowing them to stretch up towards the poles, this does not mean that meteorologists believe that the vortex actually controls the whole circulation pattern. It is far more likely that the root cause of any changes lies in the tropical region where the energy input to the weather machine is greatest. But the changes in the circumpolar vortex, especially the distinctive shift from strong to weak circulation patterns, are far easier to pick out than the subtle changes in the energy balance in the equatorial zone. What really happens is that the weather machine changes gear, and the first clear sign of this is that the zigzagging path of the westerly flow around the pole changes character. When we look for the causes of climatic change, it would be a mistake to look too long at the regions of the globe dominated by westerly wind flow. It makes more sense to look at changes in the amount of energy reaching the ground – the insolation – and at changes in the amount being reflected away again into space. The South Pole receives more insolation on a summer's day than any other place on Earth, yet it remains one of the coldest places on Earth simply because it is covered by snow and ice, so that very little of the incoming solar energy is absorbed. If snow and ice extended far enough into the temperate zone, with no change in insolation from the present-day pattern, the Earth would be in the grip of a persistent Ice Age – bare grass or forest reflects only 15 per cent of the insolation it receives, while fresh snow reflects more than 80 per cent.

Perhaps merely a succession of bad winters would suffice to tilt the climatic balance in favour of a new Ice Age. George Kukla, of the Lamont-Doherty Geological Observatory in New York, calculates that in the severe Northern Hemisphere winter of 1971–2 snow and ice cover increased by one sixth of the amount needed to start the 'next' Ice Age. Could six such winters in a row tip the balance that far? Perhaps that is overstating the delicacy of the climatic balance today, but many sober, respectable climatologists accept that a century of severe weather could be enough to initiate a new Ice Age. As we shall see in the next chapter, the key discovery is that Ice Age conditions are normal in the world today and interglacials are rare. An unusual combination of factors melted enough snow and ice between 11,000 and 18,000 years ago to start the present interglacial; since then the pattern has returned to normal and we are living at a time when insolation across the globe is characteristic of the Ice Age pattern. It's just that the snow hasn't got around to falling yet; but when it does, the insolation will be too weak to drive it away.

While we are waiting for the next Ice Age, however, smaller-scale fluctuations of climate will continue to be as important as they were in the 1970s, and more important with every year that passes with growing population and a rickety world food system. Now that I have sketched the outlines of our present understanding of the weather machine, is it really possible to figure out the implications of the odd piece of grit intruding into the cogwheels?

The neglected seas

It would help if we knew rather more than we do about the influence of a major component of the weather machine which I have largely ignored so far – the oceans. Climatic researchers agree that the oceans must play a vital role in the workings of the weather machine. After all, the atmosphere is actually in contact with water, not land, over 72 per cent of the surface of the Earth. The circulation of the atmosphere is driven by heat, but the amount of heat stored in a column of air extending from the surface of the Earth to the fringe of space is only equivalent to the amount of heat in a similar column of water extending from the surface of the sea to a depth of just three metres. The ocean, in fact, is the initial storehouse of incoming solar energy, and it is the ocean, far more than the land, which ultimately releases heat to the base of the atmosphere to drive the weather machine. At the same time the oceans provide a buffer against sudden change, because of the great size of their heat store. Just as temperate climates are moderated by the influence of the sea to the west, so the climate of the whole Earth is buffered to some extent by the moderating influence of the oceanic heat supply (although, as we shall see, in one crucial way at least the oceans may be involved in feedback processes which exaggerate the strength of one series of climatic fluctuations).

Tim Barnett, of the Scripps Institute of Oceanography in La Jolla, California, is one of the climatologists now struggling to incorporate the workings of the oceans into a satisfactory detailed picture of the global climate system. It is an uphill task, but there is already clear evidence that the ocean influences climate not only through its own circulation system of great currents such as the Gulf Stream (which carries warm water northwards in the western Atlantic and across to keep Britain temperate even though it is at the same latitude as Newfoundland) but also through more subtle influences whereby warm and cold patches of surface ocean water help to set up characteristic wind-flow patterns over a region perhaps extending across the North Pacific. These in turn reinforce the pattern of sea surface temperatures, producing a positive feedback which strengthens the local atmospheric circulation pattern and its influence on the workings of the weather machine as a whole. Sea surface temperature 'anomalies', produced by upwelling warm or cold water, provide a source of extra heat, or a 'sink' draining heat away, which must have a profound influence on the atmosphere above.

To take one example, the climate of North America depends very much, as we have seen, on the exact zigzag path taken by the circumpolar vortex, or jet stream, from year to year and in any one particular year. At the beginning of the 1950s the characteristic pattern was for the westerlies to bring relatively warm air and mild winters to the northern US. But in 1957 the pattern changed dramatically, with the route taken by the jet stream switching to produce a series of harsher winters which continued through the 1960s. With hindsight, this particular change in the zigzags of the circumpolar flow can be clearly related to changes in the pattern of warm and cold surface waters in the Pacific. The cold-winter pattern was linked to a pattern of sea surface temperatures, with waters north of Hawaii colder than the average over the entire Pacific Ocean, and warm water on the east of the ocean, down the western coast of North America. The effect of this pattern on the atmospheric circulation was to swing the depressions moving from west to east south across the cold waters of the central Pacific before they bent north around the fringe of the warm water region and then swung almost due east across northern North America. By the time the airstream reached the north-eastern states it had travelled across a great area of the cold northern part of the continent, and so the winters in the east were cold.

In 1971 the pattern changed again, with the warm and cold patches of the Pacific swapping over. (They don't really 'swap', of course; but the region of higher sea surface temperature became cold – for reasons unknown – and the cold region became warmer.) Following the same pattern of behaviour in relation to warm and cold surface waters, the jet stream – and the depressions under it – now swung north around the warm patch, then south around the cooler waters and into California before swinging away north-eastwards across the continent, carrying warm south-western air to bring mild winters to New England.

This is just one example of the way ocean temperatures affect weather and climate in just one region of the globe. But even this has much wider repercussions. Because the zigzagging jet stream must always complete an exact number of loops in its circuit of the globe, so that the head of the wriggling snake of wind always catches up with its tail, a flip of the pattern over the Pacific and North America shifts the whole pattern by a corresponding amount downstream, to the east. If the jet zigs north-eastwards across America, then by and large it has to zag south-eastwards across the Atlantic and turn north-east again to bring mild southerly air to Europe. With a weak circulation being powerfully driven by sea surface temperature patterns in the Pacific, the overall pattern requires that a run of mild winters in New England is matched by a run of mild winters in western Europe.

Of course there are other factors involved as well, so the agreement between North Pacific temperatures and the severity of winters in, say, New York and London, is far from perfect. But the evidence for the reality of such connections has now been established, and the task oceanographer/climatologists such as Tim Barnett have set themselves is that of fitting more and more pieces of the jigsaw into place until a complete picture of the influence of the oceans on climate emerges. That goal is still some way in the future: one problem, says Barnett, is that the present educational systems simply do not train scientists to be equally adept at meteorology and oceanography; another is that while meteorology is seen as a 'relevant' scientific discipline with immediate practical benefits and gets funded accordingly, oceanography is classified as a more abstract branch of research ('blue sky' research would be the usual term, but it is deliciously inappropriate in this case!) and gets much less funding. The real breakthroughs in understanding – and predicting – climate and weather are likely to come when oceanography is given the attention it deserves. Meanwhile, it is still possible to make a fair stab at climatic prediction, and the explanation of past patterns of climatic changes, by concentrating on the handful of factors which affect the insolation, the amount of solar heat actually reaching the surface of the Earth.

The key climatic influences

The first possibility is that the amount of heat arriving at the top of the atmosphere, the solar 'constant', itself varies, regardless of what happens to the radiation on its way through the atmosphere to the ground. As a rule of thumb, various calculations indicate that a one-per-cent change in the amount of heat reaching the ground would change surface temperatures by about $1°$ C, so, other things being equal, a one-per-cent change in the Sun's brightness would also be equivalent to a $1°$ C change in surface temperature. Astronomers have traditionally regarded the Sun as a stable, unvarying star – but it may come as a surprise in the space age to learn that there is no direct evidence that the solar constant really is constant to an accuracy of one or two per cent. As far as actual measurements of the Sun's output are concerned, it is quite possible that variations big enough to explain the Little Ice Age and the preceding little optimum can occur on the appropriate time-scale of decades and centuries.

It is also possible – and in the eyes of many experts looks rather more likely – that the total heat output of the Sun is constant, but the nature of the radiation changes slightly from time to time. If a little more of the Sun's radiant energy were to be produced in the ultra-violet, and a little less in the visible spectrum, then the influence of the stratosphere, the ozone layer, would be correspondingly stronger, since it is ultra-violet radiation that is absorbed in this region of the atmosphere. The same sort of effect could occur even if the Sun's output was unvarying in both quantity and quality, if other influences produced a change in the concentration of ozone in the stratosphere. More ozone would mean more solar ultra-violet energy absorbed, with less getting through to warm the ground below; less ozone in the stratosphere would allow a very modest global warming below.

Lower down, in the troposphere, weather itself can be involved in feedback loops which alter the climatic balance. If there are more shiny white clouds in the atmosphere, then more of the incoming solar radiation is reflected away into space without ever reaching the ground. One theory of Ice Ages – no longer fashionable today – took this idea to extremes by arguing that Ice Ages occur when the Sun, for whatever reason, becomes unusually hot. The idea was that the first effect of an increase in the solar constant would be to evaporate water from the oceans, producing an extensive cloud cover around the world and reflecting incoming heat away. Beneath the cloud banks, the Earth would cool dramatically, with the moisture in the clouds falling as snow. By the time the clouds dispersed the land would be covered with a white blanket of snow and ice, which could take over the job of reflecting away solar heat, and a new Ice Age would have begun. This quaint theory fell from fashion because we now have better explanations of the ebb and flow of ice, and because in any case it is rather contrived. But climatologists admit that their computer models of the changing climate do not take adequate account of changes in cloud cover, and that the balance between clouds and radiation is still very much an area that needs more investigation.

A related influence on climate comes from dust in the atmosphere, and this problem has received a lot of attention recently, inspired by Professor Bryson's claim that dust from human activities may be blocking out so much of the Sun's heat that it is hastening us into the next Ice Age. Originally, the idea stems from the historical evidence that great volcanic eruptions like Tambora in 1815 are followed by a dip in global temperatures which may last for several years. The effect is thought to be due to volcanic dust penetrating into the stratosphere and spreading a thin veil around the globe, reflecting out incoming solar heat. Mount St Helens had very little influence on global temperatures because the explosion that tore it apart blew out sideways, blasting dust over a wide area but not directly up into the stratosphere. Other volcanoes certainly have influenced climate, and Bryson believes that the 'human volcano' – dust from wind-blown soil, smoke from factory chimneys, and so on – is now a bigger influence than any natural volcano on current climatic trends. The debate continues to rage, because the influence of this kind of microscopic dust particle on the climate could go either way. In terms of reflecting heat, the question is whether the 'grey' dust is covering 'black' farmland, in which case it will be reflecting away heat that would otherwise be absorbed, as Bryson believes, or whether the grey dust layer is above a shinier surface below, perhaps snowfields, in which case by absorbing some solar energy, and some energy reflected back from the shiny ground surface below, it helps

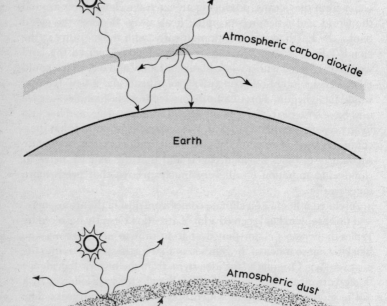

Figure 3.4. Some theorists argue that manmade dust in the air is blocking heat from the Sun and causing the Earth below to cool. Others say that the build-up of anthropogenic carbon dioxide is trapping heat that would otherwise be lost to space, causing the Earth to warm.

to warm the atmosphere. Most climatologists believe that on balance the human volcano is having very little influence on global temperatures today. That doesn't mean they are right – the consensus has been wrong before. But it is certainly true that Bryson is out on a limb with his more extreme forecasts of an imminent, anthropogenically stimulated, return of the Ice Age.

Throughout the troposphere, of course, man is undoubtedly changing the natural balance by causing a build-up of carbon dioxide. The consensus here is that this is a real problem which will lead to a noticeable 'greenhouse effect' warming of the globe within twenty years, and that problem is the subject of Part Two of this book. At the surface of the Earth mankind is producing other changes, less well publicized than the greenhouse effect, which tend to have the opposite influence on climate. Clearing darkly coloured tropical forest for farmland, which may then become an arid dust bowl without the protecting trees to maintain the ecosystem, produces an increase in the reflectiveness of the surface – its 'albedo' – which results in more solar heat being reflected away into space. This may become an important problem within another human generation; on the same sort of time-scale, equally dramatic changes in albedo might be produced in the Arctic, if Soviet plans to divert the great Yenisei and Ob rivers southwards goes ahead. Climatologists fear that taking this major source of fresh water away from the Barents and Kara seas to water the deserts of Kazakhstan and Turkmenia, 3,000 km to the south, could lead to extensive melting of the Arctic ice-cap, with unpredictable consequences for world weather patterns, especially in the region dominated by the westerly flow of the circumpolar vortex.

Knut Aagaard and L. K. Coachman of the University of Washington have sketched out a disturbing scenario of the possible consequences of a reduced flow of fresh water into the northern seas. Because salt water is denser than fresh water, the water from the rivers forms a layer on top of the salty waters of the Arctic Basin, and because the water several hundred metres deep is much saltier than the water above, it can be several degrees warmer than the top layers without convection getting to work to raise it to the surface. The increased density caused by its saltiness holds it in place against the natural tendency of warm water to rise. So the warm water stays in the depths, and the colder, fresher surface layers form an icy skin on the surface of the Arctic.

The Ob and Yenisei are big rivers. The Ob discharges an average of 400 cubic km of water each year, and the Yenisei 550 cubic km each year, which can be compared with the discharge of the greatest river on Earth, the Amazon, which is 4,800 cubic km a year. Together, the two great northern rivers amount to almost one quarter of the Amazon flow. If all of this flow were stopped (and, happily, there are no plans yet for the Soviet engineers to go that far), then a region of sea covering a million square kilometres – as big as Pakistan, or Egypt, or France and Spain combined – would be freed from ice cover as the warm waters, no longer covered by a cold lid of fresh water, rose to the surface. This would certainly upset the present climatic balance and alter the flow of the westerliés. It might – but this is merely supposition – lead to a feedback process, with the ice-free region absorbing heat from the Sun and warming still more, melting the ice next to it in a repeating pattern until the whole of the Arctic was ice-free. At present the Soviet plans are for a five-per-cent reduction in the fresh-water discharge into the Arctic Ocean by the early decades of the twenty-first century. This is roughly comparable with the range of natural fluctuations in river discharge from year to year and should not be too dramatic a departure from the present climatic balance.

Such schemes are certainly not yet in the league of the greenhouse effect as potential threats to climatic stability. But they do serve as a reminder that mankind now has the power to change the environment of our planet dramatically. Should any one country have the right to carry out projects, even within its own territory, which may have worldwide climatic repercussions? Even the Soviet climatologist E. K. Fedorov, of the USSR State Committee for Hydrometeorology and Control of the Natural Environment, commented at the 1979 World Climate Conference that 'if the ice cover of the Arctic Ocean were made to disappear, the atmospheric and oceanic circulation would adjust itself in such a way that the ice would not be able to re-establish itself'. This would, he said, 'lead to considerable changes in climate throughout the world ... it seems reasonable to suppose that specific, once-only actions could produce irreversible changes'.

Such changes might in some ways seem to be for the good. Would an ice-free Arctic really be such a bad thing? Maybe not – for the USSR, for Canada and for Norway. But the present agricultural systems of the world are fine-tuned by modern technology to make maximum use of existing climatic conditions. By tailoring both our crops and our farming techniques to a particular pattern of weather we ensure that when everything goes well yields are very high indeed, but that *any* shift away from what are now optimum conditions results in a dramatic fall in the amount of food available on world markets. This is the bind we have locked ourselves into with the farming successes of the 1960s and the Green Revolution. Anything other than 1960s weather and the world food system is in trouble, whether the change is towards warming or cooling, stronger or weaker circulation, a contraction or an expansion of the circumpolar vortex.

And there are still other natural processes which affect the climate on a time-scale important to man. One intriguing puzzle is that although over the past thousand years or more China and Japan have experienced the same pattern of little optimum and Little Ice Age conditions that have occurred in Europe and America, they seem to have experienced the pattern first! Chinese records go back to the founding of the Chou Dynasty in 1066 BC, and very detailed records of severe weather, freezing rivers, great droughts and so on are available from 200 BC. In Japan a curious climatic indicator has come down to us in records of the first blooming in spring of the cherry trees, an event of great significance to that culture. Together, the Chinese and Japanese records tell the same story, reassuring confirmation that our modern interpretation of the old records is accurate. A long cold period in the East lasted from the tenth to the fourteenth centuries, with the most severe weather, the coldest on record for the region, in the twelfth century. Can this be linked to the European Little Ice Age which set in rather later and persisted into the nineteenth century?

The answer seems to be 'yes', for the Little Ice Age did not arrive in all Europe at the same time. In European Russia the cold began to set in around 1350; in central Europe things began to get worse in the middle of the fifteenth century; and in Britain it was a hundred years later still that the Thames began to freeze almost regularly. It is at least possible that the greatest cold spell in historical times rolled from east to west right across Asia and Europe. It is simple to explain this in terms of a drift of the zigzag pattern of the circumpolar vortex, combined with the expansion of the vortex which made conditions over the whole hemisphere deteriorate. But why should the vortex drift westwards in this way? One suggestion is that it is coupled to the

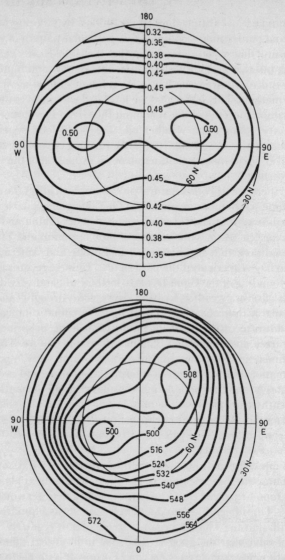

Figure 3.5. When atmospheric pressure (bottom) is plotted around the North Pole in terms of the height above sea-level at which pressure is 500 millibars, the contours bear a striking resemblance to the contours of the Earth's magnetic field (top). Is this a coincidence?

known westward drift of the Earth's magnetic field, that somehow circulation is loosely tied to the magnetic patterns. Joseph King of the Appleton Laboratory has pointed out that the Earth's magnetic field, instead of having a simple, single North Pole, actually has a double pole making a dumb-bell-shaped distortion of the magnetic field in polar regions. The circumpolar vortex, instead of sweeping neatly around the geographical North Pole, also zigzags around a double centre of high pressure over the polar regions. And the double centre of the circulation pattern sits closely over the double centre in the magnetic pattern.

Is this just a coincidence? King thinks not, and has coined the name 'magnetometeorology' for the new science of magnetic weather forecasting. Other experts are more doubtful and think it could all be a coincidence, although Lamb acknowledges that various different pieces of climatic evidence combine to show that 'some disturbance affecting the prevailing winter temperature makes steady progress westward around the [Northern] hemisphere, so as to complete one circuit in about 600 years' (Vol. 2, p. 485), and he credits King as the first person to suggest the existence of the westwardly drifting component of circumpolar vortex and the possibility of a link with the westward drift of the magnetic field. Since the mid 1970s, as we shall see in Chapter 7, the evidence for a link between magnetism and climate has grown considerably.

But the trouble with all these influences on climate is that none of them alone can explain the swings and roundabouts of climatic changes in historical times. The real world is more complicated than that, and it seems that what we have to unravel is the complex interplay of dust from volcanoes, changing magnetic fields, changes in the ozone layer, perhaps even changes in the Sun itself, and largely unknown (and as yet unknowable) changing influences from the oceans. To make a forecast of future weather, and then modify that forecast in the light of mankind's increasing output of carbon dioxide into the atmosphere, is, frankly, more of an art than a science. But the main components of the puzzle do seem to have been identified, and will be described in detail in the next few chapters. First, though, it might be reassuring to deal with one aspect of climatic change that is thoroughly understood and which can be used for reliable and accurate forecasting, even though the time-scale of the forecast may be a little too long to be of immediate practical use. Ironically, in view of the great debate about stability of climate that went on little more than a hundred years ago, we know more today about the rhythms of Ice Ages and interglacials than we do about the way the weather will shift in the next hundred years – and we can forecast the arrival of the next Ice Age with some precision.

4
The Next Ice Age

Every year the temperate regions of the globe experience the beginning of a new Ice Age. We call it winter. The fact that small changes in the distribution of the heat reaching the Earth from the Sun can produce such dramatic changes at high latitudes as the difference between summer and winter shows how delicately we are poised on the edge of an Ice Age, given the arrangement of continents which has been typical of the geography of the Earth for the past three million years. That is an important proviso, for geophysicists are now able to tell us with confidence that the present geography of the globe is just a passing phase, a temporary arrangement of continents which are in perpetual movement around the surface of our planet, breaking up, colliding to form new continents, breaking up again and sticking together again, in a stately kaleidoscope of geographical change which has gone on for at least the past thousand million years, and surely for much longer, although inevitably the geological record is more vague the further back in time we probe.

For most of the history of the Earth, and certainly for most of the past thousand million years, climatic conditions have been very different from those of the present epoch, because the geography was very different. 'Normally', there are no landmasses close to the poles of our planet, and the great ocean currents are able to circulate freely to the highest latitudes, bringing warm equatorial water to the poles and ensuring that every region on Earth, except for the peaks of high mountain ranges, is free from snow and ice. These normal conditions were those prevailing when the dinosaurs roamed the Earth and the great tropical jungles that formed present-day coal deposits flourished.

Roughly every couple of hundred million years, however, the drift of the continents carries a landmass over one or the other of the poles. Warm water can no longer penetrate to the highest latitudes, and in the cold of winter snow falls and settles on the land. In summer the white snowfields reflect away solar heat and refuse to melt; before long great ice sheets are established and an Ice Epoch has arrived. Such an Ice Epoch may last for a few million years, or even ten million years or more, depending on how long it takes for the polar continent to drift away and allow the warm waters to return. During the Ice Epoch there will be periods when the cold is more intense and periods when the icy grip eases slightly – but the glaciers never disappear entirely. Antarctica today is in the midst of such an Ice Epoch, which began sometime between 7 million and 25 million years ago.

There is another way to freeze the polar regions. Rather than one continent sitting squarely over the pole, several continents may be grouped around a central polar sea, blocking out the warm currents from the equator, so that a skin of ice develops over cold polar water. The unique feature of the present epoch in Earth history is that this kind of polar ice-cap – encouraging solar heat to be reflected away and thus allowing a build-up of ice on the land nearby – exists in the Northern Hemisphere at the same time as the more usual sort of ice-

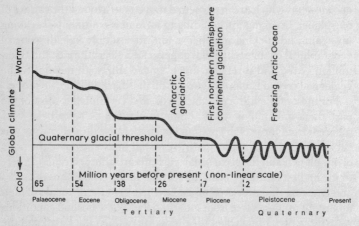

Figure 4.1a (Schematic only). Over 65 million years, since the time of the dinosaurs, the Earth has slid into an Ice Epoch, within which more or less regular fluctuations – individual Ice Ages – have occurred for about three million years. (Based on data from J. Andrews in Winters of the World, *edited by B. S. John, David & Charles, Newton Abbot, 1979. Note the non-linear scale which shows more recent events in more detail.)*

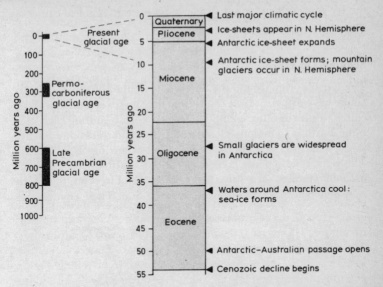

Figure 4.1b. The recent fluctuations of climate detailed in Figure 4.1a can be seen in their long-term geological context on this chart covering 1,000 million years of the Earth's history. Glacial ages like our own recur every few hundred million years as the continents shift around the globe.

cap exists in the Southern Hemisphere. We live not just in an Ice Epoch, but in a double Ice Epoch. The world began to cool about 65 million years ago, shortly after the death of the dinosaurs, but the full strength of the present ice epoch only set in about 3 million years ago when, several million years after Antarctica first became glaciated, the present northern continents slid into position, screening the Antarctic Ocean from the warm equatorial currents.

Fascinating though this story is, it is only the background to the story of the rhythm of Ice Ages within that Ice Epoch. For all practical purposes, in human terms we can regard the geography of the world as fixed, and the Ice Epoch as a permanent feature on any time-scale important to us. Why, then, is the Earth not in the grip of a full Ice Age today? Why are there ever interglacials during which the ice-caps shrink back into their polar fastnesses and the temperate regions become temperate? In a real sense the repeating rhythms of Ice Ages and interglacials are akin to the repeating rhythms of winter and summer; and to understand the astronomical causes of the Ice Age cycle, it helps to understand the astronomical causes of the seasonal cycle.

Seasonal rhythms

The Earth's orbit around the Sun is not circular, but elliptical. An ellipse has two foci, not one centre, and the Sun sits at one focus of the ellipse, while the other is empty. During its orbit – during each year – the Earth is sometimes closer to the Sun and sometimes further away, with its closest approach at present occurring on 3 January. This is the time of perihelion, when the Sun–Earth distance is about $91\frac{1}{2}$ million miles; on 4 July, at the other end of its orbit, the Earth is at aphelion, $94\frac{1}{2}$ million miles from the Sun. These changes are

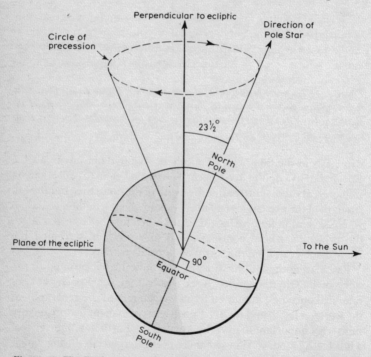

Figure 4.2. The Earth is tilted by about $23\frac{1}{2}°$ out of the perpendicular to a line joining the Earth to the Sun. Changes in this angle of tilt, and in the way the tilted Earth wobbles, or precesses, over thousands of years explain many features of climatic change during the present Ice Epoch.

significant, but small compared with the average distance from the Earth to the Sun over the whole orbit, some 93 million miles (150 million km), and they do not account for the rhythm of the seasons, as is made amply clear by the fact that Northern Hemisphere winter occurs just at the time of closest approach to the Sun! True, our northern winters are a little less cold than they would be if they occurred at the other end of our orbit around the Sun (and Southern Hemisphere winters correspondingly colder than they might otherwise be), but these changes are at a level one step more subtle than the changes which bring the march of the seasons. The true cause of the seasonal rhythms lies in the way the Earth is tilted relative to an imaginary line joining the centre of the Earth to the centre of the Sun.

This tilt amounts to about $23\frac{1}{2}°$ – that is, a similar imaginary line joining the North and South geographic poles of our planet makes an angle with the line joining Earth and Sun of $23\frac{1}{2}°$ *less* than a right angle (which is 90°). As the Earth orbits around the Sun, the direction of this tilt in space – relative to the distant stars – stays almost constant from year to year (Figure 4.3). The result is that for part of the year the Northern Hemisphere leans towards the Sun, while for the other part of the year it leans away. As seen from the surface of the Earth, the Sun climbs higher in the sky in the summer hemisphere, while the summer pole experiences 24 hours' sunshine a day in midsummer and the opposite (winter) pole experiences nights which last equally long.

When the daylit hours fill a bigger fraction of the 24-hour day, in summer, then clearly there is more solar heat available at the surface of the Earth. At the same time, because the Sun is higher in the sky its rays shine down more nearly vertically, concentrating heat upon a smaller area of the Earth's surface than when the Sun is low on the

Figure 4.3. The tilt of the Earth today also causes the rhythmic variation of climate over the year which we see as the march of the seasons.

horizon. Just as the tropics are, for this reason, warmer than the temperate zone of the Earth, so summer in the temperate zone – and at the poles – benefits from this increased insolation. The seasons are defined, strictly speaking, in terms of the orientation of the tilted Earth with respect to the Sun. When the Northern Hemisphere is tilted directly towards the Sun, which happens on 21 June, it is midsummer in the north (summer solstice) and midwinter in the Southern Hemisphere; at the opposite end of the Earth's orbit, on December 21, it is midwinter in the Northern Hemisphere (winter solstice) and midsummer in the south. Exactly halfway between these two points on the Earth's orbit we have two dates on which day and night are each exactly 12 hours long, named (from the point of view of Northern Hemisphere observers) the vernal equinox (20 March) and the autumnal equinox (22 September) (see Figure 4.4). Winter lasts from 21 December to 20 March; spring from 20 March

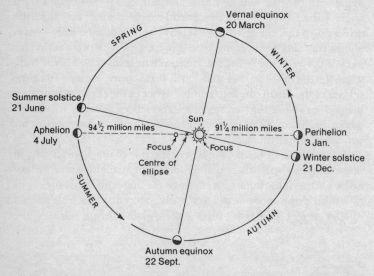

Figure 4.4. *Over the course of a year (one orbit around the Sun) the Earth's distance from the Sun also varies. Perihelion is the time of closest approach to the Sun; aphelion the point in the orbit where we are furthest from the Sun. At summer solstice the North Pole is tipped towards the Sun; at winter solstice it is tipped away. It is entirely a coincidence that we live at a time when summer solstice occurs almost at aphelion.* (Based on Figure 14 from J. and K. Imbrie, Ice Ages: Solving the Mystery, Macmillan, London/Enslow, New York, 1979.)

to 21 June; summer from 21 June to 22 September; and autumn from 22 September to 21 December. At least, it does if you are an astronomer – these dates are set solely by the geometry of the Earth's orbit and have nothing to do with the behaviour of plants and animals, or how cold you may feel! For non-astronomers, it is more sensible over most of the Northern Hemisphere to define the winter months as December, January, February and March, with a two-month Spring (April/May), a four-month summer (June to September) and a two-month autumn (October/November).

The peaks of heat in summer and the troughs of cold in winter lag behind the position of the Sun in the sky, seen from Earth, because of the time it takes for the Earth to respond to changes in the amount of insolation. In the Northern Hemisphere 22 June is not the hottest day of the year, even though it has most daylit hours, because in the weeks around that date a great deal of the incoming solar heat has to go into warming the hemisphere out of its frozen winter state. In the autumn this excess heat is only slowly dissipated away as the Sun sinks lower in the sky at noon each day, so that the shortest day, 22 December, is far from being the coldest. Thus the pattern of seasons on the ground consistently lags behind the pattern of the astronomical seasons.

It is just a coincidence that summer solstice today occurs so close to the time of aphelion, for although the dates of the solstices are almost constant over a human lifetime, the orientation of the Earth's tilt with respect to the distant stars does drift slightly over the millennia. Because of the gravitational pull of the Sun and Moon acting upon the Earth's slight equatorial bulge, the whole planet wobbles slowly and gracefully like a spinning top. The result is that the North Pole does not always 'point' in the same direction, but moves around a circle in space, so that the imaginary line joining the North and South Poles traces out a cone with an angle of twice $23\frac{1}{2}°$. It takes 26,000 years for this stately wobble to trace one circuit, and as a result the whole pattern of summer and winter solstices, vernal and autumnal equinoxes, slips around the Earth's elliptical orbit. This is called the precession of the equinoxes – though it might just as well be called the precession of the solstices – and it provided the basis for the first (albeit incorrect!) theory of the Ice Age rhythms of the past few million years.

Ice Age rhythms

Johannes Kepler, the seventeenth-century astronomer who first realized that the orbits of the planets around the Sun are indeed ellipses, not circles, also discovered that each planet traces its orbit not at a constant speed, but faster when it is moving near the part of the orbit closest to the Sun (perihelion) and slower at the end of its orbit furthest from the Sun (aphelion). The details of the variable motion are not important here, although they provided one of the bedrocks upon which Isaac Newton built his universal theory of gravitation. What does matter here is that today the half of the Earth's orbit centred on the summer solstice (Northern Hemisphere spring and summer) is travelled more slowly than the half-orbit corresponding to autumn and winter. Combined, spring and summer contain seven more days than autumn and winter; while in the Southern Hemisphere the cold seasons are, together, seven days longer than the combined warm seasons. Joseph Adhémar, a French mathematician, suggested in 1842 that this could be the reason why Antarctica is in the grip of ice today, while the Northern Hemisphere enjoys the relatively balmy conditions of what we now call an interglacial.

Because of other, more subtle shifts in the geometry of the Earth's orbit around the Sun (which we shall come on to shortly), the cycle over which the precession of the equinoxes repeats is not 26,000 but 22,000 years long at present. So, according to Adhémar's idea, 11,000 years ago, when the pattern of summers and winters was reversed, the Northern Hemisphere should have been in the midst of a full Ice Age while, presumably, Antarctica had rather less ice than it has today. That proposal fitted in very well with the limited knowledge geologists of the time had about past Ice Ages, and it fitted especially well the mass of evidence accumulating in the second half of the nineteenth century that the Northern Hemisphere had indeed suffered a succession of Ice Ages separated by interglacials. For all anyone knew in 1842, there had been an Ice Age in Europe and North America 11,000 years ago, with a previous interglacial 22,000 years ago, a previous Ice Age 33,000 years ago, and so on back into the mists of time. But ten years later, in 1852, Alexander von Humboldt, the German scientist famous for the ocean current that bears his name, pointed out the fatal flaw in Adhémar's argument. Although the Northern Hemisphere summer half of the

Earth's orbit is seven days longer than the winter half, that is because the Earth is further away from the Sun during the summer half of the orbit, and therefore receives less insolation. Averaging over the whole of spring/summer and the whole of autumn/winter the two effects cancel out – the total amount of heat received by each hemisphere over the entire year is identical.

It is appropriate that the name of Humboldt is now linked with the pattern of ocean currents around the world, since, as we have seen, the real reason why Antarctica is so cold is that ocean currents cannot penetrate to the pole. We also know now that 11,000 years ago the Earth was not in the grip of a full Ice Age, but just emerging into the present interglacial, while 22,000 years ago (when according to Adhémar there should have been an interglacial similar to present conditions) the most recent Ice Age really was at its peak (Figure 4.5). But Adhémar's idea proved of lasting importance because it started people thinking about how changes in the Earth's orbital geometry might indeed cause the ebb and flow of the great Northern Hemisphere Ice Ages. It took more than a century for this to lead to

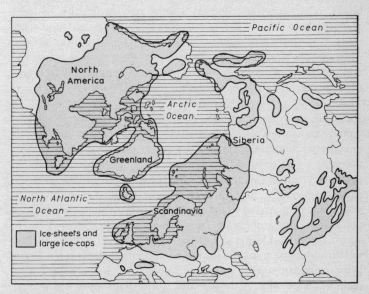

Figure 4.5. At the height of the most recent Ice Age, 20,000 years ago, the pattern of glaciation over the Northern continents looked something like this.

an accepted theory of the Ice Age/interglacial rhythms, a theory now exactly matched up with the well-understood geological record of global temperatures over the past two million years or more.

With Adhémar's false step corrected, the first steps down what turned out to be the correct path to an astronomical theory of Ice Ages were taken by a Scot, James Croll, in the 1860s. Croll's story is often told in tones of wonder as the tale of an unsung genius – the janitor at the Andersonian College and Museum in Glasgow – who rocked the scientific world with a brilliant new theory of Ice Ages. But this popular and appealing story is more than a little misleading. Croll had always been of a scientific bent and clearly of more than average intelligence, but had been unable to develop his talents through formal education because of the poverty of his family.

Born in 1821, Croll had to leave school at the age of 13 and studied science as best he could on his own while helping his mother. He became a millwright, but abandoned the trade to devote himself to his studies; he then took up carpentry but had to give that up because of a worsening childhood injury that eventually left him with a stiff elbow. John and Katherine Imbrie describe, in their book *Ice Ages*, how Croll married and opened a shop, only to find his business venture failing through lack of attention while he devoted most of his energies to study of academic books. They also detail his various itinerant jobs (including the classic one of life insurance salesman) and the failure of another venture, a hotel. Along the way he found time to complete a major book of his own, *The Philosophy of Theism*, which was well received by reviewers.

By the time the janitoring post became available, Croll was already known as a self-made scientific thinker, and he took the job deliberately because, although the salary was small, it gave him access to a fine scientific library and the opportunity to study its contents. From his cosy museum niche he began to publish scientific papers in the 1860s, and it was only natural that he should turn his attention to the mystery of the Ice Ages, one of the major talking points in science at the time. In his recent book *Ice*, Sir Fred Hoyle says that 'his work came as a considerable surprise to the academic world, and it says much for the social broadmindedness of Victorian Britain that, from 1867 onwards, James Croll was accepted as a scholar of distinction by the scientific world'. This seems a surprising comment to me – the evidence is rather that it was thanks to Victorian narrowmindedness and class distinction that the young

James Croll had to leave school at 13 and had no hope of enrolling in a university. Croll's achievements were entirely his own, owing nothing to the generosity of Victorian scientists, and this fact must have made his election as a Fellow of the Royal Society, in 1876, sweet indeed. That election followed hard on the heels of the publication of Croll's book *Climate and Time*, the summation of ten years' work on the cause of Ice Ages.

Croll's theory of Ice Ages, like Adhémar's, started from the fact of the tilt of the Earth and a knowledge of the cycle of the precession of the equinoxes. But he brought into the argument another astronomical factor. Because of the complex interplay of gravitational forces between the planets of our Solar System, the shape of the Earth's orbit changes in a regular and predictable way. The orbit is always an ellipse, but sometimes the ellipse is almost a perfect circle and sometimes it is much more elongated. These changes in the eccentricity of the orbit had been calculated by the French astronomer Urbain Leverrier (who, like Croll and all scientists of their time, had no electronic computers to aid him, remember, but had to carry out all his calculations by hand), and Croll used Leverrier's work as a platform on which to build his own theory. When the eccentricity is low, the Earth's orbit is nearly circular; when the eccentricity is high, the orbit is more elongated. The orbit constantly but gradually shifts between the two states, following a cycle roughly 100,000 years long. At present the Earth's orbit is nearly circular (eccentricity close to zero); scores of thousands of years ago it was relatively elongated (eccentricity about 6 per cent). So Croll argued that a circular orbit corresponded to the warm conditions of the interglacial, while the elongated, high-eccentricity orbit was the cause of the recent Ice Age.

But how could the effect work? Even though the shape of the Earth's orbit changes, Leverrier's calculations showed that the total amount of heat received by the whole Earth in the course of a complete year is always the same (with the unstated assumption, of course, that the heat output of the Sun is indeed constant). Croll guessed that the balance of heat between the summer and winter seasons must be important, even if the total over the whole year stayed constant. He argued that if winters are cold, then snow can accumulate more easily, and once it does so it will reflect away incoming solar radiation and keep the Earth cold. (Indeed, Croll may have been the first person to incorporate this idea of positive feedback into any scientific theory, although it is now common and crucially important to the latest ideas on Ice Ages, as we shall see.) Thus if the Earth is far from the Sun during Northern Hemisphere winters there ought to be an Ice Age. When the Earth's orbit is circular there is no such seasonal difference in insolation for this effect to work on, which seemed to explain the current interglacial. And even when the orbit is elliptical it seems that the precession of the equinoxes also plays a part in fine-tuning the weather machine to respond to the small changes in insolation over the whole stretch of the Earth's orbit.

Croll's theory conflicted with no known geological evidence of the time, and, although some scientists were doubtful how such a small change in the eccentricity of the Earth's orbit – over a range of a few per cent – could have such dramatic repercussions, it had an enormous initial impact. Debate about Croll's proposal continued in scientific circles until the end of the nineteenth century, but then, faced with mounting evidence that the most recent Ice Age had ended not scores of thousands of years ago but only about 10,000 years ago – so that the Earth had still been in the grip of ice when the orbit was already nearly circular – geologists abandoned the idea. It took a combination of still better calculations of the details of the astronomical influences and comparably accurate measurements of the ebb and flow of ice over the past hundreds of thousands of years to show eventually that the astronomical rhythms do modulate Ice Ages. The inspiration for this twentieth-century development of Croll's work came from the Yugoslav astronomer who has given his name to the astronomical theory of Ice Ages, Milutin Milankovitch.

The Milankovitch Model

Milankovitch added one other astronomical influence to the two factors considered by Croll. As well as the precession of the equinoxes and the variation of the orbital eccentricity, the angle of the Earth's tilt changes, nodding up and down between $21 \cdot 8°$ (more nearly upright) and $24 \cdot 4°$ (maximum tilt) over a cycle 41,000 years long. The present $23\frac{1}{2}°$ tilt (more accurately, $23 \cdot 4°$) is roughly halfway between the two extreme possibilities. For the past 10,000 years the tilt has been decreasing. Since it is the tilt of the Earth that produces the cycle of the seasons, this means that the differences between seasons are less extreme now than they were 10,000 years ago – other things being equal, summers are a little cooler and winters are a little warmer than they used to be. Clearly, this tilt cycle must have a strong influence on any climatic mechanisms affected by the precession cycle; and if it really is the seasonal differences in insolation that produce the rhythm of alternating Ice Ages and interglacials, then the tilt cycle cannot be ignored. This completes the tally of effects which change the orbital geometry of the Earth and alter the balance of heat received by each hemisphere in the summer and winter seasons. Combining the three effects, it is possible to calculate the amount of insolation received at any latitude on Earth in each season at any time in the past. That was an enormous task in the days before electronic computers, but Milankovitch calculated the appropriate radiation curves for a range of latitudes from 5° north to 75° north, publishing the results in 1930. His story – a saga even more impressive than that of Croll's struggle for scientific respectability – is told in detail in the Imbries' book.

Apart from the extensive calculations of radiation curves, which provided a mine of information for a generation of climatic theorists, Milankovitch's key contribution was to incorporate a suggestion from the German geologist Wladimir Köppen into the astronomical theory. This suggestion was that it is a reduction of *summer* insolation, not the occurrence of very severe winters, that is the key to Ice Ages. The argument, taken up today as a cornerstone of the modern astronomical theory of Ice Ages, is that winters in Arctic regions are always cold enough for snow to fall and glaciers to grow, but that in summer the glaciers melt – or, at least, they do today. The way to spread ice across the globe is to reduce the summer melting as much as possible, so that the re-growth of the glaciers in winter more than compensates for the summer losses.

This turned Croll's older idea on its head and reversed the pattern of warm and cold intervals which the theory 'forecasts' ('backcast' would be a better term) for the recent geological past. Working with Köppen, Milankovitch convinced himself that the pattern of insolation curves for the region of the globe north of about 60° matched up very well with the geological record of Ice Ages – cold northern summers really did coincide with Ice Age maxima. But the geological evidence was far less accurate, and less precise, than the detailed radiation curves calculated from the mathematics of the astronomical theory. In the 1940s the agreement between the two looked very good; by the end of the 1950s revisions of the geological time-scale and new interpretations of the evidence seemed to leave the Milankovitch Model out of step with the real world, and it was discarded by all but a handful of climatologists. But in the past decade a synthesis of still better theoretical calculations (at last we do have high-speed electronic machines to do the donkey work for us!) and unprecedentedly accurate measurements of the Earth's past temperature, obtained from analysis of cores of sediment drilled from the sea-bed, has shown that Milankovitch was substantially right all along – the astronomical influences do modulate the climate, and the key season is Northern Hemisphere summer.

The modern synthesis

Although many people have been involved in synthesizing the modern version of the astronomical theory of Ice Ages, two in particular can be seen as representing the two facets of Milankovitch's own work. The Belgian astronomer André Berger has produced what are probably the best and most detailed calculations of the varying seasonal insolation at different latitudes, while Czech-born George Kukla, now resident in the US, is typical of the theorists who have pinpointed *exactly* which seasonal influences are important for the spread or, as now seems most crucial, the retreat of the ice. At the time of writing, probably the most up-to-date scientific account of the cause of interglacials has appeared in a paper in *Nature* by Kukla and Berger, together with R. Lotti and J. Brown; but no matter how assiduously the theorists have beavered away to refine the Milankovitch Model, the reason it is so widely accepted today is that the record of temperature variations on Earth in the recent geological past is now known to match almost perfectly the 100,000-year, 41,000-year and 22,000-year rhythms of the three factors which change the orbital geometry of the Earth with respect to the Sun.

The measurements depend on scientists being able to obtain a supply of raw material untouched by the ravages of geological processes for hundreds of thousands of years, and on their being able to interpret the material as a thermometer. The first requirement is met by the skill geophysicists and oceanographers have developed in extracting long cores of sediment from the deep ocean bed, where layer upon layer of material builds up as the years go by. A great deal of this sedimentary material is made up of the shells of dead sea creatures, the microscopic foraminifera, which in the fullness of geological time will produce new deposits of chalk. As the sediments pile up, with the youngest deposits at the surface and older chalky shells buried deeper in the mud, different layers of sediment down a long core provide samples from different periods of time. Several techniques can be used to pinpoint the dates when particular samples were being laid down. Unfortunately the annual layers cannot be counted directly, like growth rings in trees or ice layers in cores drilled from the Greenland ice-cap, but some radioactive elements present provide an indication of age and the sediments also carry traces of changes in the Earth's magnetic field, fluctuations which have been very well dated from the study of rocks on land. The magnetic evidence, in particular, provides a 'calendar' accurate enough for the climatologists' needs.

Nobody knows for sure how or why the polarity of the Earth's magnetic field reverses from time to time, but reverse it certainly does. Studies of magnetic rocks laid down at different times in the history of the Earth show that sometimes the north and south magnetic poles are in the opposite hemisphere from their present locations. The evidence comes from old rocks, accurately dated by standard geophysical techniques, which carry a fixed record of the magnetism at the time they were laid down. Magnetic elements in a molten lava will, like tiny compass needles, line up north–south while the rock is still fluid; but once it is set they are frozen into place and cannot change their orientation even if the Earth's magnetic field changes. From then on they carry a permanent record – fossil magnetism – of the orientation of the Earth's magnetic field at the time the rocks set. It is now clear that quite often, in terms of geological times, the magnetic field dies away over a period of a few hundred or thousand years, then builds up, over the same sort of time-scale, in the opposite sense. Sometimes it dies away and then builds up again in the same sense – but that doesn't affect the argument.* The pattern of repeating magnetic changes through

* It is, perhaps, important to stress that there is no evidence whatever that the Earth's geographic poles flip over in space while the magnetic field remains stationary. Several bizarre 'theories' based on this misconception have appeared in both science fiction (which is fair enough) and books labelled as science fact, which is nonsense. The phrase 'polar reversals', in geophysics, means changes in the Earth's *magnetic* poles and does not refer to the Earth toppling over in space!

geological time provides a unique calendar against which other events can be calibrated. Deep-sea sediments contain magnetic material which aligns with the Earth's field as it is being laid down. So fossil magnetism gives climatologists a means to date the samples of foraminiferal remains in different layers of sediment in a deep-sea core.

With the dates fixed, the temperatures of long ago are determined by a variation on the same technique of isotope measurements used for the Greenland ice samples. The carbonate of the chalky material contains oxygen, in the same two isotopes – oxygen-16 and oxygen-18 – as every other sample of oxygen on Earth, and the ratio of isotopes depends on the temperature of the Earth at the time the marine creatures were alive. Another trick is to count the numbers of different species of foraminifera in different layers of sediment – some thrive in warmer water and some in cold, so the types found in each layer also give a guide to temperature. From the late 1960s onwards, ever improving measurements on ever better core samples began to fill in the outlines of the history of the Earth's temperature over tens of thousands of years. Then, in 1976, what is now regarded as the definitive study was published in the pages of *Science*.

Jim Hays, John Imbrie and Nick Shackleton reported a study based on the analysis of two deep ocean cores, one providing a good continuous record of the past 300,000 years of climate and the other, although spoiled at the top, providing a clearly readable global thermometer for the period from 100,000 to 450,000 years ago. With the overlapping region in the middle providing a check of the two cores against each other, the result was an unambiguous record of changing global temperatures going back for nearly half a million years, ample to test for the presence of even the longest, 100,000-year, Milankovitch cycle. The whole length of core covering 450,000 years is 15 metres; samples were taken at intervals of 10 cm throughout its length, representing 150 sample dates at 3,000-year intervals throughout prehistory – a short enough span to provide an accurate test for the existence of the shortest, 22,000-year, Milankovitch cycle.

By the time these core samples were analysed it was already clear from many similar investigations, and other geological evidence, that glaciations – Ice Ages – occur simultaneously in both hemispheres of the globe, at least during the present Ice Epoch. This provided a puzzle for the theorists and seemed to rule out the simple version of the astronomical theory as developed by Milankovitch, but it meant that the two cores from the Indian Ocean could be taken as representative of the whole globe. The statistical analysis of the samples, carried out by Imbrie, left no room for doubt. The dominant pulse-beat of climate revealed by the isotope test and other techniques was indeed 100,000 years long, with lesser, but still clear, cycles indicated at 42,000–43,000 years, at 23,000–24,000 years and, curiously, at 19,000 to 20,000 years. There were no other significant cyclic variations on any time-scale from 10,000 years to 100,000 years, and the agreement with the astronomical theory, although not perfect, looked good – except for that odd 19,000-year cycle. The difference between 41,000 years and 43,000 years, for example, is less than 5 per cent – certainly within the possible errors involved in the measuring techniques – and the 19,000-year cycle could, after all, be due to some other process. Nobody claimed that *only* the Milankovitch rhythms determined the pattern of temperature changes!

But when Imbrie checked his results with André Berger, he discovered an unexpected bonus. Berger, able with the aid of modern computers to do 'number crunching' on a grand scale to refine the astronomical theory, found that the precession cycle is not quite as simple as it first appears. Rather than being a single, simple 22,000-year variation, it is made up of two paired cyclic variations, a major cycle 23,000 years long and a minor cycle of 19,000 years. The core sediments had not only found the expected three Milankovitch cycles, they had also found that one of the three cycles is itself a double cycle – something the theorists had not predicted, but were able to explain after the event. This convinced even more scientists of the validity of the basic Milankovitch Model. They had still to explain, however, the physical processes which turned the varying seasonal insolation pattern into a repeating pattern of Ice Ages and interglacials.

That explanation is still not complete, but by 1981, five years after the publication of the Indian Ocean core evidence, the theorists were clearly well on the way to producing a complete account. The paper by Kukla, Berger and their colleagues summarizes the state of knowledge today, and it will be surprising if there are any major changes in this picture, the modern synthesis itself. The synthesis starts from the crucial position that the *normal* state of the Earth today is a full Ice Age, thanks to the present arrangement of the continents, and that the role of the astronomical theory is, therefore, not to explain why Ice Ages occur, but why they are interspersed with short-lived interglacials. Kukla's team used a variety of indicators of past climate, including the oxygen isotope technique, to build up a comprehensive picture of global temperature fluctuations over the past 350,000 years, and then they looked at the astronomical rhythms to identify the characteristic pattern of these effects that precedes the onset of an interglacial. Even within an Ice Age there are variations of temperature and the ice advances and retreats in step with the astronomical rhythms: warm periods tend to occur when perihelion is in September, while the coldest phases of the Ice Age fluctuations culminate when perihelion is in March. But full interglacials, rather than a mere easing of the worst Ice Age conditions, are much rarer. In the past 350,000 years there have been only four interglacials, which started 335,000, 220,000, 127,000 and 11,000 years ago. The exclusive characteristic that links these recent interglacials is that they began when there was a June perihelion *and* the tilt of the Earth – its obliquity – was greater than $23.8°$. Without a June perihelion, the tilt effect alone is not enough to cause a major retreat of the ice; without the tilt effect, even a June perihelion cannot bring enough summer warmth to melt the glaciers. But when the tilt is large – so that the difference between the seasons is pronounced – and the perihelion is in June, also bringing extra warm summers and extra cool winters, the two effects together can just melt enough ice for a temporary alleviation of Ice Age conditions.

With this evidence before us, it is clear that Milankovitch was right on one all-important point. Interglacials are indeed produced when maximum summer heating is available to melt the Northern Hemisphere ice. Summer is the key season. Why does the Southern Hemisphere follow the same pattern? One possible explanation is that, fortuitously, just the opposite conditions are needed to encourage ice to grow in the south. Whereas cool northern summers allow snow that falls on land around the Arctic to stay unmelted and build up into ice sheets, in the Southern Hemisphere there is no ice-free land near the pole even today, and snow which falls into the sea will melt regardless of the insolation, or lack of it. To encourage a spread of sea-ice in the south we need very severe southern winters, so that sea-water will freeze. And since cold southern winters go hand in hand with cold northern summers, the curious geography of the globe today may be responsible for producing Ice Ages in step in both hemispheres.

The alternative way of looking at the puzzle is to say that the south is always in an Ice Age, and that only the Northern Hemisphere really changes much at all because of the seasonal insolation changes, so that the whole global pulse-beat of temperature, indicated by the isotope thermometer, is driven by the north. According to that picture, the south does cool and warm with the same rhythm as the north, but directly because of the way changes in the Northern Hemisphere affect the temperature of the whole globe and the circulation patterns of the great ocean currents. This is one aspect of the Ice Age problem that still provides plenty for the theorists to puzzle over. As we shall shortly see, the most promising line of attack on these remaining aspects of the puzzle comes from improved understanding of the interactions between oceans and ice cover. First, however, it is only right to acknowledge that a few theorists are still not persuaded by the evidence in support of the Milankovitch Model in its modern form.

A dissenting voice

The voice of dissent has recently been heard most loudly from Sir Fred Hoyle, the English astrophysicist. In his book *Ice* Hoyle dismisses the Milankovitch effects as being too small, in terms of the energy involved in the insolation changes, to account for the size of the temperature fluctuations of the Earth during the present Ice Epoch. He accepts that the Milankovitch rhythms show up in the analyses of isotopes from ocean cores, and in other data, but believes that this is simply due to the minor modulating influence of the astronomical rhythms on the existing ice-sheets. The climatologists would go along with that – they know full well that the ultimate cause of the present Ice Epoch is the arrangement of continents on the surface of the Earth, and that the Milankovitch cycles are, relatively speaking, ripples on the surface of the Ice Epoch. Where Hoyle diverges from the conventional wisdom, however, is in his view that not even all the Milankovitch cycles pulling together can haul the Earth out of the full Ice Age and into a temporary interglacial state. Instead, he prefers the idea that the impact of giant meteorites with the Earth can cause both the sudden onset of full Ice Age conditions, and the sudden arrival of interglacial conditions.

This is such a fascinating idea that it deserves to be right, and I can hardly do justice here to a theory Hoyle propounds in a whole book.

Very simply, however, he argues that the impact of a large stone meteorite with the Earth can cause a cooling by spreading dust through the atmosphere, like a mighty volcanic eruption, and blocking out the heat from the Sun. He correctly points out that the oceans of the world store great quantities of heat energy, and reasons that an Ice Age will only begin if the dust layer persists long enough for this heat to be dissipated as the Earth cools below. At a critical temperature, the atmosphere itself would become so cold that tiny water droplets suspended in the stratosphere would be frozen into ice particles, highly reflective 'diamond dust', and this would then take over the task of reflecting away the solar heat after the meteorific dust settled. To end such an Ice Age, an equally dramatic event is called for to melt the diamond dust. Hoyle favours the possibility that a metallic meteorite might strike the Earth, spreading not dust but a layer of conducting material through the upper atmosphere. Conducting, metallic material would absorb solar heat, raising the temperature sufficiently to melt the diamond dust and initiate an interglacial.

There are, says Hoyle, about ten times as many stony meteors in the Solar System as there are metallic meteors, so by and large Ice Ages ought to be ten times more common than interglacials. This argument smacks of sleight of hand, however, since it does not account for the rather even spacing of interglacials, each some 10,000 years long, separated by Ice Ages, each some 100,000 years long. Nor does the theory offer any hope of explaining – except as a coincidence – the evidence from some studies that, while climates of the past million years have been dominated by the three Milankovitch cycles outlined here, the pattern between 1 and 2 million years ago was for the coldest phases of the Ice Ages to be less intense, and for the dominant pulse-beat of climates to come from the 41,000-year rhythm, not the 100,000-year cycle. This is just the sort of challenge that might be met by a combination of the Milankovitch Model and the theory of continental drift, perhaps providing new insights into both fields of study. All the meteorite theory can say is that the pattern is just one of those things.

But giant meteorites most definitely have hit the Earth, and quite possibly could initiate an Ice Age, even if the continents were in the wrong places for a proper Ice Epoch, through spreading dust in the stratosphere. Indeed, a respectable school of thought holds that just such an event may have been responsible for the death of the dinosaurs some 65 million years ago. It seems rather unlikely, though, that two different kinds of meteorite would conspire to strike the Earth, not just in a regularly repeating pattern, but in a pattern which shows up, to climatologists on Earth, exactly in step with the Milankovitch rhythms! Hoyle's theory simply will not work as an explanation of fluctuations on the scale of the glacial/interglacial transitions – it is too drastic and too rare a process, as well as being too erratic to fit the known interglacial timetable. I am sure that Hoyle is wrong to propose that meteorite impacts can explain the regular pattern of recent climatic fluctuations, not just the big one-off events like the death of the dinosaurs. But I am equally sure that he is right to stress the role of the oceans in determining the pattern of major climatic fluctuations – the occurrence of interglacials – during the present Ice Epoch. Ironically, a report describing a combined ocean-feedback/Milankovitch model of interglacials, and pointing the way for further work on this aspect of the problem, was published in *Science* in May 1981, just a month before Hoyle's entertaining but misleading book was published.

The role of the oceans

Yet again this study came from a team at the Lamont-Doherty Geological Observatory, clearly the world centre for the continuing development of the modern astronomical theory of Ice Ages. This time the researchers involved were William Ruddiman and Andrew McIntyre, and the period they chose to look at in detail was the past 250,000 years. They started out from the evidence of the isotope record that the insolation curves for latitudes 45°N and 65°N reproduce the pattern of ice fluctuations revealed by the isotopic record (which might just possibly be a coincidence) and tried to find a physical reason for the connection. Since most of the world's heat is stored in the oceans, that is where they, like Hoyle, looked for the driving force – and they seem to have found it.

Leaving to one side the details of the calculations made by Ruddiman and McIntyre, their explanation for the pattern of Ice Ages and interglacials, and the probable feedback role of the North Atlantic ocean, can be simply explained. We know from the isotope studies and other evidence that when ice is growing rapidly, summers are relatively cold. But the oceans take some time to cool off after the warmth of an interglacial, so at this phase of the climate cycle, when the Earth has begun to move into a cold winter phase (for the Northern Hemisphere), there will be an ample reserve of oceanic warmth to keep the Atlantic ice-free. This provides a lavish supply of moisture, water evaporating from the warm surface of the ocean, which will fall as snow over the land at high latitudes during the cold winters, rapidly building up ice-sheets and hastening the Earth back into a full Ice Age. Ice Ages are linked with cool summers which cannot melt the developing ice sheets; cool summers go hand in hand with relatively warm winters, and the winter warmth is a bonus which ensures efficient evaporation of the ocean water, providing an additional boost to the process by which warm water is evaporated, falls as snow and builds up into glaciers. Eventually the ocean cools and sea-ice extends across its northern reaches; but by then the world is in the grip of an Ice Age, and the high albedo of the ice itself helps to maintain the frigid conditions, even with much less evaporation and a greatly reduced snowfall. So how does the Ice Age end?

At a time of rapid deglaciation, the orbital pattern brings unusually strong summer insolation over the ice-covered latitudes of the Northern Hemisphere. This is balanced by the intensity of winter cold. The first step in the initiation of an interglacial comes, according to the calculations of Ruddiman and McIntyre, with an increase in the amount of calving from glaciers in the hot summers, producing a spread of icebergs and, as the ice melts, a rise in sea-level which forces sea-water under the grounded edges of ice-sheets where the land meets the sea, causing the ice to break up still further and produce even more icebergs. This is exactly what happened from about 16,000 years ago to 13,000 years ago, at the end of the latest Ice Age. The biggest effect of the iceberg influx was between 40° N and 55° N in the North Atlantic, where the top 100 metres of ocean water must have been cooled by several degrees during the initial deglaciation. So the interglacial we live in began, like earlier ones, with warm summers and cold oceans still covered by large areas of sea-ice. These two factors combined to minimize the amount of moisture evaporating from the sea – no moisture can evaporate from under a lid of ice, and very little will evaporate even from ice-free water if the temperature of the water is close to freezing. So, although the winters at the time were cold, there was very little moisture available to make clouds and produce snowfall which could rebuild the glaciers which were suffering such extensive summer losses. The ice cover also has an effect on our old friend the circumpolar vortex. The westerly storm track tends to follow the edge of the sea-ice under glacial conditions, and this helps to keep what storms there are, and the moisture they contain, to the south of the shrinking ice-sheets.

This model explains neatly why it is a combination of the 100,000-year and 23,000-year cycles that dominates the pattern of recent Ice Age fluctuations, and it explains the rather surprising discovery, from the isotope thermometer, that the 100,000-year rhythm is the dominant feature. During each 100,000-year cycle following an interglacial the ice continues to build up, even though the build-up suffers setbacks at intervals of 23,000 years and 41,000 years. Each successive build-up is bigger than the last and more than compensates for the ice lost during each setback. But the bigger the ice-sheet is, the more efficient are the feedback processes which help to make it disintegrate at the appropriate stage of the 23,000-year precession cycle. Even a strong precession influence does not cause a rapid deglaciation when there is only a modest ice-sheet around in the first place, paradoxical though that seems at first. But, once the extent of the ice-sheet reaches some critical value, the next time the precession cycle brings a June perihelion and a warm summer, disintegration is inevitable. The end of an Ice Age – the termination – always occurs at the time of the first precessional maximum after a large volume of ice has built up. In round terms, it takes four precessional cycles for enough ice to build up, and four lots of 23,000 years is close to 100,000 years. The surprising strength of the 100,000-year cycle in the global temperature record is explained – what the isotopes are actually measuring is a small, genuine 100,000-year effect, combined with a bigger pseudo-100,000-year rhythm which is really the product of an enhancement of the 23,000-year cycle after sufficient build-up of the ice-sheets.

These calculations and the study of Kukla's team which suggests an important role for the angle of tilt do not entirely mesh together, which is hardly surprising since they are both so recent. The theorists always like to explain everything with their own pet theory of the moment and only grudgingly accommodate other ideas into an improved overall picture of how things really are as the years go by and the evidence mounts up. Plenty of work remains to be done before it is time to dot the i's and cross the t's of the definitive astronomical theory of Ice Ages, but the work so far leaves no room for doubt that the Milankovitch rhythms do modulate the present

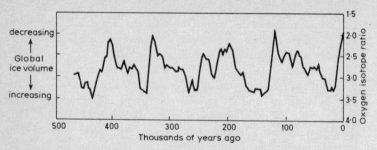

Figure 4.6a. Data from the cores analysed by Jim Hays and his colleagues (see Bibliography) reveal the pattern of temperature changes on Earth over the past half million years. On this evidence, we live in a short warm period ('interglacial') which must soon end, bringing back full Ice Age conditions, if nature has her way.

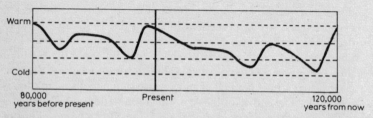

Figure 4.6b. Calculations of the Milankovitch rhythms made by Nigel Calder put more detail into the present pattern of climatic change. It seems that the present interglacial is already past its peak. The forecast is downhill all the way into the next Ice Age.

Ice Epoch and explain the occurrence of interglacials. No room for doubt, that is, unless like Hoyle you favour a theory so radically different that you have to ignore all the evidence in support of ideas like those of Ruddiman and McIntyre in order to make room for it!

The next Ice Age

So when will the next Ice Age begin? Just at the moment, we are living off the accumulated benefits of the ageing interglacial. From season to season and year to year the oceans today actually help to reduce the extremes of temperature over much of the temperate zone – warm water nearby keeps us comfortable in winter, while cool water nearby stops us overheating in summer. This is an example of negative feedback, a damping down of extremes; it is natural, perhaps, that until recently many climatologists gave the same kind of negative role to the oceans when attempting to construct theories to explain Ice Age/interglacial fluctuations. But they were wrong. Today's warm oceans are, in the longer term, a menacing threat, offering a burden of moisture to the atmosphere that could, at the right time, tip the balance to cover the northern continents in a blanket of snow that would persist for a hundred thousand years. Ruddiman and McIntyre are vague on the exact trigger for the start of a new Ice Age, although they explain in great detail the timing of the start of an interglacial. Perhaps this is where the work of Kukla, Berger and their other colleagues at Lamont really comes into its own. They point out that a high tilt angle is just the thing to encourage disintegration of sea-ice, provided the precession cycle is at the right phase, but they also look ahead, comparing the pattern of insolation changes that can be calculated for the critical latitudes for the immediate future with the pattern that has been characteristic of the end of the four most recent preceding interglacials. Their conclusion is clear. The warmest times of the present interglacial are indeed past and we are moving rapidly into an orbital configuration appropriate for a full Ice Age. The natural trend of temperatures is sharply downwards from now until the first cold minimum of the next Ice Age, which will occur in about 4,000 years' time. It is impossible to predict just when, between now and that cold minimum the ice-sheets will return in force, but with the tilt angle of the Earth already down to 23.4° and decreasing, so that the difference between the seasons is being reduced, and aphelion now occurring in July, Northern Hemisphere summers are already cool enough for the ice-sheets to remain if once they become established.

It isn't just that 'normal' conditions in the present epoch are those of an Ice Age and we live in an interglacial; in terms of the orbital geometry, the interglacial is already over, but we just don't happen to have had a run of severe winters marked enough to establish the ice-sheets once again. It could have happened in the Little Ice Age; it might not happen for a couple of thousand years. It could happen next century; it will certainly have happened within the next 4,000 years. After that, the world will remain in an Ice Age for a further 110,000 years before the next interglacial.

Can this conclusion be of any practical value? It tells us one very important thing. The warmth of the peak years of the interglacial, the climatic optimum, has passed not just temporarily but for more than 100,000 years. If the Earth is indeed already on the slope down into the next Ice Age, we can expect no more climatic optimums, nor even little climatic optimums, but we can expect more little ice ages, on the way to the big one. If the big one does come within a hundred years, then all bets are off and there may be little mankind can do except learn to live with it. If, however, the big one holds off for a couple of thousand years, which is quite likely, then we should be planning for normal weather no better than the normal weather of the past millennium. Our rule of thumb that the weather of the past millennium may be our best guide to the weather of the next millennium turns out to need only a minor modification: judging by the success so far of the Milankovitch Model in explaining the occurrence and extent of interglacials, the weather of the past millennium, including the centuries of the Little Ice Age, was probably rather *better* than the weather nature has in store for us for the next millennium. On a time-scale of decades and centuries, however, that still leaves scope for variability on a scale too small to notice compared with the transition from an interglacial to an Ice Age, but of vital importance to mankind. Against the background of a steady long-term natural deterioration of climate, we can now put those temporary flickers into perspective.

5
The Solar Connection

The weather machine is driven by heat from the Sun, and its workings must be affected by any change in the amount of solar heat arriving at the surface of the Earth. The Milankovitch Model shows how important seasonal rearrangements of the Sun's heat can be, even if the average over the whole year does not change. But what if the average insolation, the solar heat arriving at the Earth's surface, does change? There is no doubt that the Sun varies in some obvious ways. Sometimes it is relatively quiet and its surface is featureless; at other times it produces great bursts of flaring activity, storms on the Sun, and its visible disc is scattered with dark blobs called sunspots. For a hundred years, astronomers have known that the storms and spots come and go over a cycle roughly 11 years long, from minimum (quiet Sun) to maximum (stormy Sun) and back again. And for a hundred years people have speculated that these changes in the Sun's activity might affect the weather on Earth. The best evidence today is that although the Sun's overall brightness varies by no more than a fraction of a percentage point over its more or less regular cycle of activity, both individual flare bursts and the average level of flaring activity over a whole solar cycle can affect the weather on Earth. On a slightly longer time-scale, the average level of storminess of the Sun may be linked with changes in the atmosphere of the Earth which are responsible for little ice ages and little climatic optima – not because the amount of heat put out by the Sun varies, but because particles streaming out from the Sun across space affect the transparency of the Earth's atmosphere.

Historically, it was the fascination of the 11-year sunspot rhythm that first led astronomers and meteorologists to speculate about the possible links between the Sun's activity and the weather. But it makes more sense to look first at the effects of individual solar storms on the atmosphere of the Earth, then work up to the influences which extend over several solar cycles. The evidence is clear-cut and very well established in articles which have appeared in scientific journals over the past thirty years or so, but it is surprisingly little known outside those specialist scientific circles.

Pioneering work by Walter Orr Roberts, now at the Aspen Institute in Colorado, and his colleagues first drew the attention of the scientific community to the evidence that storms at high latitudes on Earth – specifically those crossing into Alaska from the Pacific – are more stormy than average shortly after a burst of flaring activity on the Sun. When the Sun produces a great storm flare, electrically charged particles – chiefly protons – sweep across interplanetary space and past the Earth. Some of them, trapped by the Earth's magnetic field, spiral down towards the magnetic poles, where they produce bright auroral displays, lights in the sky produced naturally by the same process we see in operation, tamed, in a neon tube. The charged particles also affect the measured magnetic field of the Earth – the geomagnetic field. And the observations, made over a generation, show that there are changes in storm systems developing on Earth, and in the measured surface pressure variations at sites in north-west Europe and North America, two or three days after the bursts of auroral and geomagnetic activity that signal the arrival of bursts of particles from the Sun.

It is no surprise to find the effects concentrated at the latitudes of Alaska, the northern contiguous United States and north-west Europe, since the charged solar particles are concentrated in just this region by the Earth's magnetic field; and, as we have seen, these are just the latitudes where modest disturbances, whatever their origin, may have a profound effect on climate. It *is* a problem to explain just how the chain of cause and effect works. There is no really satisfactory theory to account for the observed fact that solar protons arriving in bursts at the top of the atmosphere do change the weather, at least in one part of the world, a few days later. The fact that they do, however, is an all-important introduction to the otherwise surprising discovery that the average level of solar activity – the storminess of the Sun from year to year and decade to decade – affects not just the weather for a day or two, but the climate of the globe – the 'average weather' – on a scale of a human lifetime or longer.

Sun and weather

For most of the hundred years or so since the Victorian fascination with sunspot cycles, claims that these changes in the Sun affect the weather on Earth have run up against the same problem – no known mechanism could explain the tie-up between sunspots and the weather. That doesn't seem to have worried the Victorians, who happily claimed evidence for correlations between the Sun's 11-year cycle and phenomena on Earth as diverse as the changing water-levels of lakes in Africa, crops yields in England and the fluctuations of the stock market! Twentieth-century 'discoveries' of the Sun's influences on life on Earth have sometimes been equally exotic, though generally regarded in scientific circles as slightly disreputable. But in the 1970s ever improving statistical tests, coupled with the use of high-speed electronic computers, have proved that there are 11-year and other rhythms present in weather data which can only be explained by the influence of the Sun's changing pattern of activity. The effects are small, but measurable. At the same time, astronomers puzzling over the cause of the Sun's cycle of activity, and seeking to probe the nature of the Sun's interior with observations of particles called neutrinos, have found new puzzles which suggest that we may not understand how the Sun works as well as we thought we did. The question of why the Sun's level of activity changes, and the mystery of solar neutrinos, have to be glossed over here.*

The important point is that, with some of their complacency shattered, some climatologists and astronomers have looked again, with open minds, at the possible correlations between changes in the level of the Sun's activity and the climate on Earth. This time the

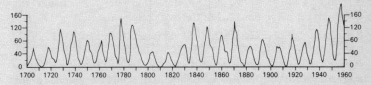

Figure 5.1. The changing level of solar activity since 1700, measured in terms of sunspot number. When the Sun is quiet, as in the early nineteenth century, the Earth is colder.

* But are dealt with at length in my book *The Death of the Sun*.

results are unambiguous – when the Sun is quiet, the Earth is cold.

The basis of these studies is the observed pattern of sunspot activity, but no one suggests that the spots themselves affect weather on Earth. Rather, we know from present-day observations that more spots on the Sun are a sign of increased flare activity, with more particles streaming across space in the so-called solar wind, and that an absence of spots implies a quiet Sun and weak solar wind. So when the historical records reveal, as they do, that during the most intense phase of the Little Ice Age in Europe, from 1645 to 1715, there were scarcely any spots on the Sun, and little trace of the 11-year cycle of activity, it gives us pause for thought.

Jumping off from this observation, Stephen Schneider and Clifford Mass, of the US National Center for Atmospheric Research, compared the changing level of sunspot activity since the seventeenth century (unfortunately the detailed observations don't go back beyond AD 1600) with the changing temperature of the Earth. They ignored the year-to-year variations, but looked at the average 'strength' of the sunspot activity in each cycle compared with the temperature of the Earth below, and they found a very good correlation between the ups and downs of temperature and of the sunspot number. The sunspot number is the astronomers' standard measure of solar activity – it is a number based on the area of the

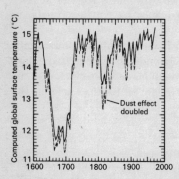 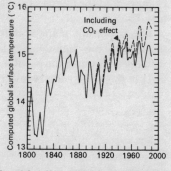

Figure 5.2. Using a computer model which allowed for effects of both volcanic dust and solar variations, Stephen Schneider and Clifford Mass were able to 'predict' a pattern of temperature changes for the Earth since 1600. The computer model agrees rather closely with temperature changes in the real world. But adding in the predicted carbon dioxide 'greenhouse effect' for the most recent period suggests that all these natural processes may soon be overwhelmed as a result of human activities.

Sun's disc covered by sunspots, but a sunspot number of, say, 50 does not mean that literally 50 spots are visible. A high sunspot number means that the Sun is more active, involving a more frequent occurrence of large flares and a more powerful, gusty solar wind. Schneider and Mass found that temperature variations on Earth over the past three centuries can be neatly explained if the amount of heat reaching the surface of the Earth changes in line with the sunspot number. The effect is as if the Sun's heat output is 2 per cent lower when there are no sunspots than for sunspot numbers between about 80 and 100, and returns to the level for no sunspots by the time the sunspot number reaches 200 or so, which is just about the highest it ever gets in record-breaking cycles of solar activity (Figure 5.2).

The 'as if' is an important caveat, for Schneider and Mass were dealing primarily with measurements at the surface of the Earth, or in the lower atmosphere. It could be that the Sun's heat output varied by 2 per cent – though astronomers were horrified at the suggestion – or it could be that something connected with the Sun's cycle of activity changed the transparency of the Earth's atmosphere, affecting the amount of heat getting through to the ground. At the time Schneider and Mass made their calculation, in the mid 1970s, a few balloon flights had been carried out in which instruments were lifted into the stratosphere 25 or 30 km above the ground in an attempt to measure the Sun's heat output – the solar constant – accurately. These did show hints of a variation about the right size to explain the climatic changes of the past 300 years, but the evidence was inconclusive, since even at these altitudes the instruments were still within the Earth's atmosphere. Since 1975, instruments mounted on satellites such as the Nimbus series have found the solar constant to be, as the astronomers predicted, very nearly constant. So far, the observations have been carried out only over half of one sunspot cycle. They are hardly the definitive last word. But the best evidence suggests that on a time-scale important to mankind the Sun's heat output varies by no more than a few tenths of one per cent, and that the link between Sun and weather has to be sought in changes in the stratosphere itself: more of this shortly.

In the years since Schneider and Mass reported their findings the spotlight has shifted back to studies which add to the growing weight of evidence that year-to-year variations in solar activity, as well as the average over a whole sunspot cycle, do indeed produce a measurable influence on the weather. In most cases the effects are far too small to be of value in everyday terms – there is no point in planning your holiday according to the phase of the sunspot cycle, since other influences are bigger and likely to disrupt your best-laid plans. But the small-scale fluctuations from year to year help to flesh out the reality of the links between Sun and Earth.

Patterns and periodicities

In most parts of the world, tree-rings provide the longest sequences of 'climatic' data that are available for statistical analysis using modern processing techniques to unravel 'signals' with low amplitude in short records of highly variable data. But only rings from some trees, in some parts of the world, can be used to trace patterns of past climatic variations beyond the span of the historical record. The first studies of this kind depended on measurements of the thickness of individual tree-rings – fatter rings denote good years for the trees, an arboreal climatic optimum. The question is, what is 'good' climate for a tree? For some species, rain is the key factor; for others, temperature. And in either case the technique lacked precision. Today, dendroclimatologists have more precise ways to estimate past temperatures, using a variation on the oxygen isotope technique that has proved so successful in analysing cores from ice-sheets. The proportions of isotopes such as oxygen-16 and oxygen-18, and of hydrogen and deuterium, reveal the temperature of the water being taken up by the tree each year, because molecules containing the heavier isotopes evaporate less easily. First, though, the dendro-climatologists have to establish that in living trees the isotopes do not migrate across the tree-rings, garbling the temperature record. This can only be done by calibrating the outer rings of living trees against the historical record of temperature variations before using the inner rings to find out temperature changes in the more distant past. And, among other things, that means throwing out the data from the famous bristlecone pines in the White Mountains of California, where there are no decent records of modern temperature changes to establish the calibration.

Nevertheless, there are trees from Europe and Japan, in particular, which have been accurately calibrated and provide a good guide to temperature changes over the past couple of hundred years. If tree-rings have been good thermometers for two centuries, they should be reliable further back in time. Data from a sequence of

Japanese cedar spanning 2,000 years and German oak spanning 1,000 years (each pieced together from different, overlapping, segments of wood) have now been analysed by Leona M. Libby and colleagues at the University of California at Los Angeles, and they show a significant long-term deterioration of climate, a cooling of about 1·6° C at the Japanese site over the past 2,000 years. A long-term cooling of 10° C, remember, would establish Ice Age conditions. But superimposed on this trend there are fluctuations of temperature from year to year, and an adaptation of the signal-processing techniques which electronic and communications engineers use to unravel their signals from background noise reveals several significant patterns.

For this particular analysis, each wood sample used covered five tree-rings, so there was no hope of picking out any 11-year solar influence. Libby's team did, however, find other patterns which have also been linked with solar activity, including a 179-year cycle over which the sunspot cycle itself seems to wax and wane, and other periods also known from the sunspot record. The 11-year rhythm is only the most obvious feature of a complex pattern of solar activity variations. The same periodicities are found in analyses of isotope data from sediments on the bottom of the Santa Barbara Basin, off California; and in a special study of a 72-year sequence of rings from the giant sequoia, rings wide enough to be analysed one by one, a member of Libby's team found the, by now expected, 11-year pattern. This study did not, however, find one other pattern that might have been expected.

During each 11-year sunspot cycle, the Sun maintains a magnetic field oriented in one direction, like the Earth's magnetic field. But at the end of each sunspot cycle the Sun's magnetic field fades away, then builds up in the opposite sense, with north and south magnetic poles reversed, as the next sunspot cycle builds up. The result is that there is a cycle of solar *magnetic* variability some 22 years long, the 'double sunspot' cycle. Perhaps this cycle too could affect the weather and climate on Earth. But there is no evidence for it in the isotope record laid down in the tree-rings.

This would only be of passing interest if it were not for the fact that droughts in the American Midwest recur at roughly 20-year intervals, and have done so since at least the beginning of the nineteenth century. Some forecasters have 'explained' this by saying that the Great Plains weather responds to the double sunspot cycle

– although in all honesty the correlation between the drought cycle and the Sun's magnetic cycle is far from exact. The tree-ring studies seem to knock that idea on the head. So how can we explain the drought cycle? The best explanation on offer came, by chance, from a mathematician using yet another technique to unravel the links between Sun and weather.

A lunar influence?

Bob Currie is, at the time of writing, an unemployed statistical mathematician. He is unemployed because although his work in recent years, on the payroll of a major US oil company, should have been directed towards discovering new reserves of oil and gas (using his statistical skills to interpret the shock-wave signals obtained by seismologists studying the geological structure of the Earth), he preferred to use his skills to unravel climatic patterns and predict future trends. His employers tolerated this for a while, even granting him leave of absence to work for a time at the NASA Goddard Space Flight Center at Greenbelt in Maryland; but in 1981 they told Currie to give up the climate studies and return to his proper duties or quit. He chose to quit, and to make what efforts he could to publicize his fears that a major drought, perhaps comparable to the dust-bowl of the 1930s, could hit the US Midwest in 1991.

This is not the forecast of a mystic or wild man of science. Currie's impeccable credentials, and the established use of the processing technique called Maximum Entropy Spectrum Analysis (MESA) in such practical work as locating oil deposits, leave no room to doubt that the patterns he finds in the temperature record are real. Very few of these patterns are surprising to anyone who has watched the unfolding saga of the links between solar activity and the weather. The surprise is that in Currie's hands MESA is such a powerful tool that it does not need 2,000 years of tree-rings to operate on, but can cope with the direct measurements of the meteorological record, spanning no more than a century or two. The historical record of measured changes in temperature and pressure across the United States is sufficient for Currie to work on, and he finds that the 11-year sunspot 'signal' is unambiguously present in temperature data from weather stations in the US – but only in data from those located east of the Rockies and north of about 35° N. When the data from all weather stations across the continent are averaged together, the solar signal is masked by other variations, and this seems to be one reason why earlier statistical studies missed it.

Why should the region east of the Rockies and north of 35° be particularly susceptible to the solar influence on weather? As Currie points out, the affected region exactly corresponds to the region where the storm tracks of prevailing westerly winds sweep across the continent, the local manifestation of our old friend the circumpolar vortex. What the evidence is really telling us is that the circumpolar vortex is affected by changes in the Sun's activity, even over individual sunspot cycles. How much more susceptible might the vortex be to a change in solar activity like the prolonged minimum which lasted for more than half a century and 'coincided' with the Little Ice Age? Currie's detection of the small short-term effect, only about 0·18° C variation in temperature in the affected region over one sunspot cycle of activity, points the finger at bigger variations affecting larger regions of the globe on longer time-scales. So much Currie might have expected to find from his analysis of the meteorological record. But he certainly did not expect to find an 18½-year rhythm, affecting the weather of the US Midwest but corresponding to no known solar cycle. Instead, this particular rhythm seems to fit closely another astronomical rhythm, the 18·6-year interval between peak tides raised by the Moon in the atmosphere and oceans of the Earth, the lunar nodal tides.

The tidal pattern arises because peak tides occur when the Earth, Moon and Sun are *precisely* aligned. Because the orbit of the Moon around the Earth is tilted slightly compared with the orbit of the Earth around the Sun, this does not happen every month but only every 18·6 years. For the same reason the alignment is also highly important in calculating eclipses; this leads to the whimsical speculation that Stonehenge, the 4,000-year-old monument in southern England that is certainly a good eclipse-predictor, could be used, if the fancy took us, to predict weather changes in the US Midwest. Currie has, however, given us a better forecast using his rather more modern computers.

Once again the measured effect is small, and once again the affected region lies downwind from the Rockies in the prevailing west-to-east air-flow. Currie suggests that the effect operates because the tug of the Moon, operating with an insistent 18·6-year rhythm, can tilt the balance between the main flows of the westerlies travelling either north or south around the Rockies, with profound effects on the rainfall to the east. He has not yet worked out details for such a mechanism – nor has anyone else, and he might be barking up the wrong tree. But the reason for his concern at the implications lies with the apparent agreement between this newly discovered 18·6-year rhythm in weather and the 20-year Midwest drought cycle.

Table 2: Droughts, sunspots and lunar cycles

Lunar cycle	Drought years	Sunspot cycle
1805	1801	1798
1824	1823	1823
1843	1845	1843
1861	1862	1867
1880	1882	1889
1899	1900	*
1917	1917	1913
1936	1935	1934
1954	1955	1953
1973	1975/6	1976
1991	?	1998?

* No corresponding minimum

The dates of actual drought years in the US Midwest fit more closely the dates of peak lunar nodal tides than they do the dates of the 'double sunspot' cycle

That '20-year' cycle in fact fits the 18·6-year rhythm much better than it does the 22-year double sunspot cycle, as Table 2 shows. Since 1801, the peak drought year in the US Midwest has never been more than two years out of step with the maximum lunar nodal tide, which can be calculated accurately from the Moon's orbital parameters. The double sunspot cycle matches up almost as well, but with one highly significant run of exceptions. In 1882 there was a severe drought exactly seven years 'early' by the double sunspot cycle; in 1900 there was a drought in the middle of the cycle, halfway between 'expected' droughts; and in 1917 the drought was four years 'late'. This is exactly the pattern of behaviour we would expect if the 22-year and 18·6-year cycles were fortuitously almost in step throughout the nineteenth century, passed through a shuffling phase change – a change of step – over two cycles at the turn of the century, and have since been fortuitously almost in step again for four more cycles.

With such long cycles it takes the best part of a century for the differences to become noticeable again. But now the testing time is at hand. If Currie is right, and the lunar nodal tide does provide the driving force for Great Plains drought, the next drought is due in 1991. If the sunspot influence is really the driving force, the drought won't hit until the second half of the 1990s. Either way, the picture is a grim one when we recall how much the hungry world now depends on American grain – food from the bread-basket of the world, the US Midwest. Even that possibility, however, represents no more than a minor, short-term fluctuation of the weather. We have still to explain how the changing level of solar activity can, presumably through the way it changes the transparency of the stratosphere, affect the climate of the Earth for decades at a time.

A role for ozone?

The amount of heat reaching the surface of the Earth depends not only on the amount arriving at the top of the atmosphere but also on the transparency – or transmissivity – of the atmosphere. Apart from clouds, and volcanic dust (of which more in Chapter 6), the molecules of the atmosphere itself absorb some of the incoming solar energy. The stratosphere in particular absorbs energy, through photochemical reactions involving ozone. This makes it an inversion layer, trapping the weather systems of the world in the troposphere below. The key to an influence of solar variability on weather lies not so much in the amount of heat arriving at the top of the atmosphere as in the amount of heat getting through the stratosphere to the top of the troposphere, the weather layer. And since the stratosphere is the ozone layer, changes in ozone concentration might provide a clue to changes in the transmissivity of the stratosphere.

The troposphere extends up to about 15 km above sea-level, and the stratosphere from there up to about 50 km. So even high-flying weather balloons, at about 30 km altitude, have only penetrated halfway through the stratosphere. Some concern has been expressed in recent years that some human activities, including the use of high-flying jet aircraft and the release of chlorofluorocarbons from spray cans, might destroy, or at least damage, the ozone layer. Since most of the solar radiation absorbed by the ozone layer is in the ultra-violet band of the electromagnetic spectrum, that would let more ultra-violet through to the surface of the Earth, where it would be harmful to life. Whatever the pros and cons of that debate, it is now clear that natural variations in the ozone concentration of the strato-sphere over the solar cycle and on shorter time-scales are much bigger than any likely influence from human activities, and these natural fluctuations may play a part – though probably not the dominant part – in determining how the amount of heat getting through to the troposphere varies.

A key study, reported by Ronald Angione and his colleagues in 1976, drew on measurements of solar radiation reaching the ground at different wavelengths at sites in southern California and northern Chile during the first half of this century. Because the ozone layer absorbs ultra-violet radiation, the amount of ultra-violet getting through to the ground is a good measure of the ozone concentration of the stratosphere above. Over much of the electromagnetic spec-trum, temperature changes in the stratosphere also affect the absorp-tion, which complicates the issue. But in a band from 0·5 to 0·7 micrometers wavelength, called the Chappuis Band, the tempera-ture effect happens to be tiny and can be ignored. The data studied by Angione's team – originally gathered by the Smithsonian Astro-physical Observatory – provide a clear-cut and impressive indica-tion of ozone-layer changes during the present century. At both sites the total ozone concentrations indicated by the measured intensity of radiation in the Chappuis Band vary by as much as 20 to 30 per cent, from month to month, year to year and decade to decade. The Chappuis Band absorption itself accounts for something close to 2 per cent of the total energy reaching the stratosphere from the Sun, so this measured variability can alone account for fluctuations in the amount of solar energy reaching the ground – or the troposphere – of up to half of one per cent (25 per cent of 2 per cent). This is not enough to explain the difference between present-day conditions and those of, for example, the Little Ice Age. But the discovery highlights the distinction between the amount of solar energy arriv-ing at the top of the atmosphere – the astronomical solar constant – and the amount reaching the troposphere – the meteorological solar constant.

The next step in pinning down the link between Sun and weather involves finding a reason for those ozone fluctuations and identifying other photochemical reactions in the stratosphere that absorb incoming solar energy. With one of the unpredictable twists that seem to be characteristic of the study of the climate, and with which we ought by now to be becoming familiar, what looks like the answer to the puzzle has come from a study not of the Sun's influence on weather but of the effect on the stratosphere of atmospheric nuclear bomb tests.

The missing link

Some of the balloon measurements that provided back-up for the calculations made by Schneider and Mass in their study of climatic changes since the Little Ice Age came from a pair of Soviet researchers, K. Y. Kondratyev of the Main Geophysical Observatory in Leningrad, and G. A. Nikolsky of Leningrad University. Their evidence, from balloon flights carried out in the 1960s, showed an increase in the measured amount of heat from the Sun, the meteorological solar constant, coinciding with the period over which the Sun's activity was increasing during that solar cycle, from 1964 to 1969. A decade ago they interpreted this as evidence that the Sun's heat output itself changed by a couple of per cent over the solar cycle. But in 1979 Western meteorologists were startled by a new claim from Kondratyev and Nikolsky, that nuclear weapons testing in the 1950s and early 1960s had had a pronounced effect on the stratosphere and on the weather of the world, leading up to record-breaking severe winters in 1962–3 and 1963–4. The changes in measured solar heat after 1964, they now argued, were due to the stratosphere slowly recovering from the damage caused by the atmospheric bomb tests – and the changes provided an important clue to how solar activity can also disrupt the ozone layer, explaining the link between solar activity and the weather.

When I reported this claim in *New Scientist*, it was met with good-humoured disbelief by meteorologists, who conveyed their opinion to me in no uncertain terms. But the *New Scientist* report also led to a lengthy correspondence with Kondratyev and Nikolsky, who provided me with a detailed account of their work, which at that time (mid 1981) had yet to be published in English. Moving on from the puzzle of possible anthropogenic influence on weather, operating through the effect on the stratosphere of atmospheric nuclear tests, they make the proposal that a very similar effect, related to the changing intensity of cosmic rays penetrating the atmosphere, could provide the long-sought mechanism to explain how the Sun's roughly 11-year cycle of activity – and longer periods of quiet Sun like that of the late seventeenth century – could influence the weather on Earth. This work has not yet been through the mill of scientific peer review and formal publication in the scientific journals; it still has rough edges and can be taken as no more than an outline guide to the way the mechanism *might* work. Even so, it is important enough to rate a mention here, with the warning that the details, at least, may be changed as a result of the discussions among the experts that are certain to ensue when the work is formally published.

We can get some idea of the complexity of the puzzle from the way ozone concentration in the stratosphere changed during the 1960s, while the meteorological solar constant was increasing. Total ozone concentration over some North American observing stations increased by between 4 and 8 per cent during the 1960s and has since declined slightly, as Linwood B. Callis, of the NASA Langley Research Center, and his colleagues reported in *Science* in 1979. The Langley team interpreted this as due to changes in the Sun's ultraviolet output, the possibility that the quality of solar radiation (in a sense, its colour) rather than the quantity might vary over the solar cycle. The changes do, however, fit rather neatly into the scheme proposed by the Russian researchers, as long as we can understand how *increased* ozone can be linked with *increased* solar heat getting through to the gound. Far from the Chappuis Band absorption telling the whole story, it seems to be overwhelmed by some other effect acting in the opposite sense.

The explanation is simple enough. Quite apart from the ultraviolet absorbed by the ozone itself, solar energy at different wavelengths is absorbed during photochemical reactions which destroy ozone. If the stratosphere is flooded with some molecular compounds in particular, then they absorb solar energy and destroy ozone; the Earth below cools, while the ozone concentration of the stratosphere is reduced. When the polluting molecules are removed, the ground below warms while the stratosphere recovers – exactly as observed in the 1960s. What are the likely polluting molecules? Thanks to the scare about the effects of exhaust products from high-flying jets, many calculations were made during the 1970s on the effect of oxides of nitrogen, collectively dubbed NO_x, on the stratosphere. They exactly fit the bill – and they are produced in copious quantities when atmospheric nitrogen and oxygen combine in the fierce heat of the atomic fireball of a nuclear explosion, which rises high in the atmosphere and spreads its pollutant oxides of nitrogen widely through the stratosphere. The Russians' idea is by no means as crazy as it seems at first sight – those old wives' tales that 'the bomb' could be blamed for bad weather have more than a grain of truth in them after all.

A bomb surprise

The fireball from an atmospheric nuclear explosion reaches a height of 30–45 km above the ground. For each TNT Megatonne (Mt) equivalent of the explosion, about 3,000 tonnes of NO_x is produced, some 10^{32} (1 followed by 32 zeros) molecules. In their haste to explode as many bombs as possible during the run-up to the partial test-ban treaty of 1963, the superpowers let off an accumulated equivalent of 340 Mt of nuclear explosions in a couple of years, releasing $1\frac{1}{2}$ Mt of NO_x into the stratosphere, spreading at heights between 20 and 50 km. In the case of a typical constituent of the oxide brew, nitrogen dioxide (NO_2), the lifetime of the molecules from each explosion in the stratosphere is typically about four years, so that as well as this burst of NO_x production in 1963 the stratosphere still carried a burden of pollution from the series of atomic tests that had gone on right through the 1950s. Altogether, according to the calculations made by Kondratyev and Nikolsky, pollution equivalent to the explosion of 980 Mt was present in the stratosphere at the beginning of 1963. Assuming all of this pollution was spread through a ring of the stratosphere between latitudes 25° and 85° N, they found that each square centimetre of the globe below that ring had a burden of 10^{17} (1 followed by 17 zeros) molecules of NO_x above it, sufficient to reduce the flux of solar radiation at balloon altitudes – essentially, at the top of the troposphere – by $2\frac{1}{2}$ per cent. This fits in very well with the available measurements of solar radiation from the balloon flights of the 1960s.

It also fits in very well with the pattern of weather in the Northern Hemisphere in the early 1960s. Between 1958 and 1964, the Northern Hemisphere outside the tropical zone – almost exactly the region underneath the nuclear pollution cloud – cooled by about half a degree. The winter of 1962–3 was memorably cold in Europe, one of the two worst winters in living memory, and although the winter of 1963–4 was less severe near the centres of population, and so holds no similar place in folklore, the average over the whole hemisphere was even colder than in 1962–3. 'Beginning with 1963,' says the Soviet team, 'the frequency of negative temperature anomalies increased over the whole globe. Maximum values of the temperature drop were registered in 1964–65–66.' But while the Earth cooled at ground-level – and throughout the troposphere – the stratosphere warmed. Rocket-borne instruments showed that in the equatorial zone of the middle stratosphere the temperature in 1963–4 exceeded the average for the next 12 years by no less than 6° C. In the following years, temperatures in the stratosphere fell at all latitudes in the zone from 46–55 km altitude, and, at a more gradual rate, at altitudes between 16 and 45 km. The increased temperature of the stratosphere indicates that it is absorbing more of the incoming solar energy; the warm stratosphere is a direct cause of the cool troposphere below, because it sequesters heat that would otherwise get through to warm the ground.

From 1963 to 1967 the stratosphere slowly decontaminated after the build-up of NO_x from almost 20 years of atmospheric nuclear explosions. Between 1967 and 1970, however, a further series of bomb tests by the French and Chinese delayed the return to normal

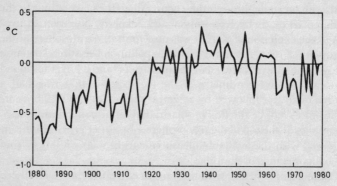

Figure 5.3. The pattern of temperature variations in the Northern Hemisphere over the past 100 years, measured as annual deviations from an arbitrarily chosen baseline. Was the dip in temperatures in the 1960s related to atmospheric testing of nuclear weapons?

conditions, and this is reflected in the changing ozone concentration reported by observing stations around the world, including those mentioned by Callis and his colleagues. Overall, the evidence in support of the idea of a link between nuclear tests and changes in the weather is too strong to dismiss lightly; however, Kondratyev and Nikolsky are the first to point out that other effects have also played a part in determining the weather patterns of recent years.

In Chapter 6, the effects of dust and other material from volcanic eruptions on the weather and climate will be investigated in more detail. But it doesn't need a genius to realize that dust carried into the stratosphere, from whatever source, could also absorb incoming solar heat, warming the stratosphere and cooling the Earth below. It happens that there was a major volcanic eruption, of Mt Agung in Bali, on 30 March 1963. Agung has been widely blamed for the pattern of climatic changes in the 1960s, but the Russians argue that because Bali lies south of the equator (at $8°S$), and because the eruption occurred late in the Northern Hemisphere winter, the pattern of atmospheric circulation would have carried no more than 15 per cent of the dust it threw into the stratosphere north of the meteorological equator. Taking this into account, they find that the Mt Agung eruption can explain no more than half of the actual cooling in the early 1960s – a cooling which had, in any case, already begun before the mountain erupted. They argue that volcanic and bomb NO_x effects 'almost equally contributed to' the change in intensity of solar radiation reaching the troposphere, 'and apparently produced similar climatic consequences'. That point we can perhaps leave for debate among the experts. Kondratyev and Nikolsky's extension of their theory to explain the connection between solar activity and the weather brings us back, however, to the main theme of the present chapter.

The cosmic connection

The basic argument seems to be that anything mankind can do, nature can do better. Molecules of the oxides of nitrogen are produced naturally in the atmosphere as a result of interactions involving cosmic rays, including those charged particles from the Sun that produce bright aurorae, disrupt the magnetic field, and in some unknown fashion stimulate the growth of storms at some latitudes. The more cosmic rays there are, the more NO_x there is, and the less ozone. When cosmic ray intensity builds up, the stratosphere should warm, while the troposphere below, where we live, cools. Because the Sun produces more particles – solar cosmic rays – when it is more active, it might seem at first sight that *increased* solar activity should always lead to a *decrease* in surface temperatures on Earth. But things aren't quite that simple. Before we move on to the complexities, however, let's look at the size of the forces involved.

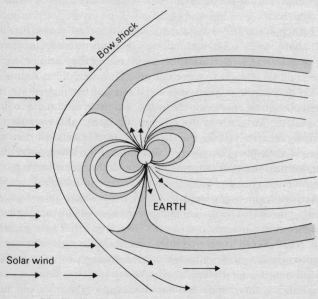

Figure 5.4. The Earth and its magnetic field are embedded in the wind of particles blowing past from the Sun. A stronger solar wind protects the Earth from cosmic rays from beyond the Solar System – but a very strong solar wind produces its own disruptive influence as particles funnel in towards the poles down the magnetic lines of least resistance.

Paul Crutzen, G. C. Reid and their colleagues, working in Boulder, Colorado, made some calculations along these lines in the mid 1970s. They showed that following a very large solar flare, called a solar proton event, as much NO_x is generated in the stratosphere at high latitudes (above 60°) as by a 50-Mt nuclear explosion. The Boulder team was particularly interested, at the time, in the implications of such changes in the stratosphere for life on the surface of the

Earth. A bigger than normal solar flare, especially if it occurred during a time when the Earth's magnetic field was relatively weak and its shielding effect reduced, could knock out so much ozone that ultraviolet radiation would flood through to the troposphere almost unimpeded. If that happened, plants might die off in profusion, while the animals that fed on them starved. Occasional 'mass extinctions' of species are known from the fossil record; could they be due to a combination of weak magnetic field and an extra strong burst of solar activity?

Whether or not that speculation is correct, there is nothing speculative about the calculation that a rare – but by no means unique – solar proton event equivalent, in terms of its effect on the stratosphere, to a 50-Mt hydrogen bomb would increase the concentration of NO_x two or three times above the natural level, decrease the ozone concentration at high latitudes by 25 per cent and produce effects lasting at least two years before the atmosphere recovered. When the Sun is very active and intense bursts of solar protons are produced, then the effect must be a cooling of the troposphere. But when the Sun is quiet or only moderately active another factor has to be considered.

Apart from the Sun's varying activity, the whole Solar System, including the Earth, is bathed by a more or less steady background flow of cosmic particles from the Galaxy at large. These galactic cosmic rays (GCR) are equivalent, in their effect on the stratosphere, to about half the influence of a major solar proton event. But their influence is ever present, not short-lived like that of a solar flare. And galactic cosmic rays interact with the atmosphere in a different way from solar cosmic rays (SCR).

SCR reach only high latitudes, guided by the Earth's magnetic field, and they are absorbed at high altitudes, in the stratosphere. GCR penetrate the shielding magnetic field and the atmosphere more easily, so that they are absorbed in the lower atmosphere at altitudes between 11 and 15 km. Also, although their intensity in the latitude band from 50° N to 50° S is only a little more than half the intensity at high latitudes, they do penetrate the equatorial region where solar cosmic rays are conspicuous by their absence.

So the overall influence of cosmic rays on the stratosphere depends on the combined influence of SCR and GCR. When the Sun is quiet, galactic cosmic rays dominate, NO_x is produced and carried into the stratosphere, and the Earth below cools while the stratosphere warms. As the Sun's activity increases, because the Earth's magnetic field is a good shield against solar cosmic rays the first noticeable effect is that the solar wind, blowing out from the Sun more strongly, deflects galactic cosmic rays and shields the Earth from their influence. The result is that less NO_x is produced, and the stratosphere cools while the troposphere warms. When and if the Sun's activity rises still further, however, the solar cosmic rays produced in the bigger flare bursts are powerful enough to penetrate the stratosphere effectively, building up NO_x once again and causing the Earth's surface to cool.

This is exactly what the Russian team found, and had some trouble explaining, in their original analysis of balloon flight data, back in 1970. The meteorological solar constant, the amount of heat from the Sun getting through to the troposphere, increases as the Sun's activity, measured by sunspot number, increases. But for sunspot numbers just above 70, the increase stops, and for still higher sunspot numbers the pattern is reversed, so that for very high sunspot numbers, up around 200 or so, the measured meteorological solar constant is the same as for very low sunspot numbers. The effect is as if the Sun's temperature first increased and then decreased with increasing sunspot number, over a range of about 2 per cent. But the cause is the changing transparency of the stratosphere, due to cosmic ray interactions.

Why has it taken so long to notice the effect – why, indeed, is it not obvious to us on the ground that the meteorological solar constant varies as sunspot number varies? One reason is that the Sun does not always reach the same levels of activity in every solar cycle. One cycle may indeed peak out at sunspot numbers of 180 or so, showing the whole pattern of behaviour outlined above. Another may reach peak sunspot numbers of only 50 or 60, producing what seems at first sight a very different pattern of temperature changes on Earth over an 11-year period. And, besides, the weather machine cannot really respond to these changes on a year-to-year basis. When the amount of incoming energy increases, its most important climatic effect is to warm the surface of the oceans slightly – but that takes years to produce a significant effect, and within five years or so the solar effect will already be operating in the other direction again. The oceans act as a buffer, smearing out any short-term influence on climate, so that all we can really expect to see are longer-term averages, over one or several solar cycles. That is why the cold of the Little Ice Age stands out so clearly as an indicator of the links between Sun and weather; in the middle of the twentieth century the Sun has been much more active, and the overall effect has been to shield us from galactic cosmic rays and allow a slight global warming, compared with the nineteenth century.

Which way will it jump next? Here we really are entering into the realms of speculation and guesswork. But one man who has managed to second-guess the Sun/weather relationship better than most in the past thirty years is Hurd C. Willett of the Massachusetts Institute of Technology. He has a good track record – but he makes gloomy forecasts for the 1980s and beyond.

Another chilly forecast

Willett was using the evidence of a link between solar activity and the weather to predict climatic changes long before anyone had worked out a plausible mechanism for the Sun/weather link. For most of the past 30 years his forecasts have flown in the face of the perceived wisdom of the day – and they have invariably proved right. Like all good studies of long-term climatic fluctuations, his work is based upon northerly and southerly shifts in the zonal wind system which sweeps around at temperate latitudes, shifts which, for simplicity, can be regarded as producing either a high latitude zonal (HLZ) or low latitude zonal (LLZ) state. The more the circulation pattern shifts towards one of these idealized states, the more the weather changes from what we regard as 'normal'. Pronounced LLZ circulation corresponds to cold and wet conditions and 'coincides' with periods of reduced sunspot activity such as the Little Ice Age, while HLZ circulation brings warmer, drier weather to the US and Europe and is associated historically with decades when solar activity reached moderate overall levels.

Critics of Willett's approach to long-range forecasting have long dismissed all this as coincidence, and it is true that he has only 250 years of completely reliable sunspot data on which to base his calculations and from which to extrapolate into the future. But the evidence for the Sun/weather link is now more secure than ever before, removing the main force of the critics' arguments. It does, however, still take a degree of scientific chutzpah for Willett to use just two and a half centuries of data to come up with a suggestion that the Sun's basic 11-year rhythm is itself modulated by two great cycles, 80 and 100 years long, which follow one another and together produce the appearance of a 180-year rhythm. The fact is, though, that MESA techniques like those used by Bob Currie have in recent years also picked out these long-term cycles, providing statistical support for Willett's intuitive reading of the data. Willett has achieved remarkable successes with his resulting forecasts. But there is a snag.

First, the successes. In 1951, when most of his colleagues were worrying about global warming, then recently detected, in the first wave of concern about the greenhouse effect, Willett correctly predicted a significant fall in temperature over the 15 years leading up to 1965. In 1955 he predicted that the increased hurricane frequency then causing concern had already reached a peak which would not be repeated in the 1960s, and that the drought then threatening the US Great Plains would shortly break. He was correct again, as he was with other predictions for the 1970s. All of these forecasts were based on the 80- and 100-year patterns of sunspot variation, coupled with the then *ad hoc* rules about the relation between Sun and weather gained more or less by practical experience. However, the pattern Willett has been using broke in 1980, with the Sun seemingly changing gear. In one way this is bad news; but in another it is a boon. Willett has made forecasts for the decade ahead assuming that the 80- and 100-year patterns will still be followed. If these are borne out, but the Sun's pattern of behaviour remains out of step, it will prove that Willett was just lucky before and the Sun's influence on weather is much less than he thought. On the other hand, if the Sun and weather stay in step, even though the 180-year rhythm has changed, that will confirm the validity of the Sun/weather link. Neither forecast, however, is very reassuring.

In the mid 1970s Willett presented his forecast for the decades ahead on the assumption that the 180-year rhythm would continue as before. He foresaw decades of low solar activity to the end of the twentieth century, a period reminiscent of the early part of the nineteenth century, which was a time of low sunspot numbers (see Figure 5.1, p. 111) and the last great wave (so far) of the Little Ice Age. The forecast for the 1980s and 1990s was for intensification of the LLZ pattern, with cold and wet conditions setting in across the middle latitudes, including the US and Europe.

But the Sun refused to play ball. Instead, 1979–80 saw a very high peak of sunspot activity, quite unlike anything that happened 180 or 360 years ago, producing sunspot numbers in the 180s in the autumn of 1979, slipping back a little, then rising again to 179 in May 1980, and still in the 150s by April 1981. The provisional mean sunspot number for the whole of 1980, released by the Sunspot Index Data Center in Brussels in the spring of 1981, was 155, higher than the peak of 105 reached in the previous maximum of 1968–9 and not far off the all-time record of 190 reached in 1957. ('All time' here means the past 300 years!) Willett now surmises that, for reasons unknown, the Sun may now be returning to a state of high overall activity which in the past has been typical of the second half of each of the two known '100-year' cycles, rather than staying at the low level of activity 'aprioriate' for the beginning of a new '80-year' cycle. If the Sun stays active, Willett expects that the cold LLZ circulation will remain for much of the 1980s, as the weather systems lag behind solar changes, but that a return to warmer, drier conditions will set in in the 1990s. If the sunspot cycle 'takes off on a new tangent', says Willett, then the 'basis of long-term climatic prediction is lost'. But 'at least if the pattern of climatic change is observed to relate to the new sequence of solar change in the manner analogous to that of the past, then the hypothesis of the solar control of climatic change is verified'.

In the light of the recent work by Currie, Kondratyev and Nikolsky, the tree-ring researchers and others it will be surprising indeed if the patterns of climatic change and of solar variability do now go their separate ways after marching in step since the Little Ice Age. All things considered, though, Willett's detailed forecasts should be treated with caution. The best guide to future weather we can glean from the solar connection comes from taking a slightly longer-term view. Compared with recent centuries, the Sun has been unusually active in the middle part of the twentieth century, and that has at least contributed to the warmth of the world during the past human lifetime. Unless some real and fundamental change is taking place inside the Sun – in which case all bets are off – the most likely forecast for the half-century ahead is for a return to normal: fewer sunspots even in peak years than in the peak years of the past few decades, and an associated global cooling. The same common-sense approach we applied to the historical record – expect what we've had before – gives us, when applied to the solar connection, the same unwelcome forecast. A cooler world than the one we are used to is normal. And it doesn't look as if the volcanic dust effect is going to change the forecast much.

6
Fire and Ice

Probably the first person to suspect a relationship between volcanic eruptions and climatic changes was the American polymath Benjamin Franklin. In 1784 he was living in Paris as the first diplomatic representative and plenipotentiary of the newly formed United States of America, and wrote:

> During several of the summer months of the year 1783, when the effects of the Sun's rays to heat the Earth in these northern regions should have been the greatest, there existed a constant fog over all Europe and great part of North America. This fog was of a permanent nature; it was dry and the rays of the sun seemed to have little effect towards dissipating it, as they easily do a moist fog . . . They were indeed rendered so faint in passing through it that when collected in the focus of a burning glass, they would scarcely kindle brown paper. Of course, their summer effect of heating the Earth was exceedingly diminished.
>
> Hence, the surface was early frozen.
>
> Hence, the first snows remained on it, unmelted . . .
>
> Hence, perhaps the winter of 1783–84 was more severe than any that happened for many years.
>
> The cause of this universal fog is not yet ascertained. Whether it was adventitious to the Earth, and merely a smoke proceeding from consumption by fire . . . or whether it was the vast quantity of smoke, long continuing to issue during the summer from Hekla, in Iceland, and from that other volcano which arose out of the sea near the island, which smoke might be spread by various winds over the northern part of the world . . .
>
> It seems, however, worthy of inquiry whether other hard winters,

recorded in history, were preceded by similar permanent and widely extended summer fogs . . .*

Many other climatologists have little doubt that the severe winter Ben Franklin experienced in Paris in the 1780s really was caused, in large measure, by the eruptions in Iceland. With almost 200 years of observations of both volcanoes and the weather since 1784, the correlation is clear enough, although volcanoes alone cannot explain all the ups and downs of the climate. And Franklin's plea for a study of the historical record to ascertain the reality of the link between volcanic eruptions and climatic changes has been answered, as recently as 1970, by a massive historical survey carried out by Hubert Lamb, then working at the UK Meteorological Office. Lamb's meticulous compilation records all volcanic eruptions from 1500 to the 1960s, and relates their impact on the atmosphere of the Earth to a standard set by defining the influence of the 1883 eruption of Krakatoa as 1,000 units on a 'Dust Veil Index'.

Between Benjamin Franklin's speculations and Hubert Lamb's Dust Veil Index, however, lie several generations of scientific theorizing and lay concern about the effects of volcanoes on the weather. The first dramatic landmark of the nineteenth century was the eruption of Tambora in Indonesia, in 1815. We now know, from historical and geological evidence, that this volcano blasted three times as much dust into the atmosphere as did Krakatoa, seven decades later, but communications were poor in the early nineteenth century, Europe was more interested in the defeat of Napoleon than natural disasters on the other side of the world, and there was no Benjamin Franklin around to draw a connection between the eruption in Indonesia and the appalling summer suffered by Europe and North America in 1816. This now well-established connection is the jumping-off point for a great deal of modern work, including Reid Bryson's studies reported in *Climates of Hunger*.

Before Tambora erupted there had been a series of lesser volcanic eruptions around the globe, between 1811 and 1813, so that the high atmosphere – the stratosphere – was already burdened with dust before the Indonesian volcano blew its top. The accumulated effect was that 1816 became famous in Europe as 'the year without a summer', while the more earthy American pioneers described it as

* Quotation from Franklin's letter in H. H. Lamb, *Philosophical Transactions of the Royal Society*, Vol. A266, pp. 425–533, 1970.

'eighteen hundred and starve to death'. Throughout the middle latitudes of the Northern Hemisphere, average temperatures were about 1° C cooler than the long-term average for the time, and measurements from some parts of England indicate a summer averaging a full 3° C below normal. Across the Atlantic, New England and eastern parts of Canada suffered equally severely, with snow falling over a wide area in June 1816 and frosts occurring in every month of that year. Crops died in the fields on both sides of the Atlantic; the Irish suffered severely, and there were food riots in Wales. All of this emphasizes not just how susceptible the weather machine seems to be to volcanic eruptions, but also how susceptible agriculture is to a fall in summer temperatures of a couple of degrees. Harsh winters may bring problems of their own, but they don't damage crops; cool summers can bring havoc to the world food markets. The implications of a modern Tambora would be many times worse than the ravages suffered by the world in 1816; and the 1980 eruptions of Mount St Helens (of which more shortly) came close to producing another year without a summer.

Impressive though the effects of a single volcanic eruption may be, however, what really matters is the total burden of dust in the stratosphere, and its influence on the weather machine. The most extreme speculations about the averaged-out effects of many volcanic eruptions occurring around the world in a short space of time come from those climatologists who argue that dust in the atmosphere alone can cause the onset of an Ice Age. Their argument is rather weakened by the fact that the Milankovitch rhythms explain the pattern of recent Ice Age fluctuations so well, but there is some geological evidence that times of increased ice cover over the globe were also times of increased volcanic activity – the dust from the volcanoes, when it does settle, leaves a detectable layer in the rocks which can be dated reasonably well. There is, however, an explanation of this other than the obvious one that the volcanic dust caused the ice cover to increase by blocking out heat from the Sun. Perhaps the increased ice cover, pressing down on the rocks of the continents, caused the increase in volcanic activity, squeezing magma out of volcanic reservoirs like toothpaste from a tube. This is by no means a facetious argument, and it fits the facts at least as well as the idea that volcanoes trigger Ice Ages. This is one argument that will continue to rage, with no prospect in sight of a definite answer one way or the other. Happily, though, this is one academic debate which has no relevance to the way climate changes on a time-scale important to mankind.

Krakatoa and the global dust veil

Krakatoa is the classic example of a volcanic eruption that is important in human considerations, as its place of eminence in Lamb's Dust Veil Index (DVI) indicates. It was also the eruption that set the American meteorologist Harry Wexler thinking about the links between volcanoes and the weather back in the middle of

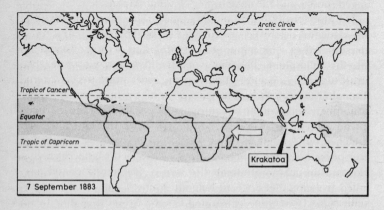

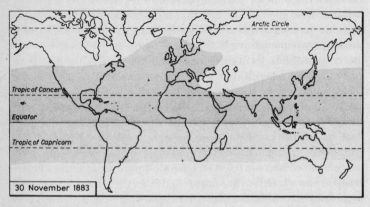

Figure 6.1. The Royal Society report on the Krakatoa eruption detailed the spread of the volcanic ash around the globe at high altitudes.

the twentieth century, and it was his work that laid the foundations of our present understanding of those links.

By the 1880s, with improved communications, and no long-running wars to distract them, scientists and non-scientists alike were able to marvel over the changes in the atmosphere produced by the eruption of Krakatoa. A special Royal Society report on the eruption included a section 312 pages long under the heading 'On the unusual optical phenomena of the atmosphere, 1883–6, including twilight effects, coronal appearances, sky haze, coloured suns and moons, etc.', and many ordinary people marvelled at spectacular green-tinged sunsets and the appearance of a blue Moon.

The story really begins on 27 August 1883, when the island of Krakatoa was destroyed in a volcanic explosion so vast that it threw an estimated 13 cubic miles (some 55 cubic km) of rock, dust and ash into the atmosphere. The plume of ash and dust may have reached an altitude of 80 km and certainly spread great quantities of material through the stratosphere. Carried westwards by the prevailing upper atmosphere winds of the equatorial zone, the cloud of fine particles spread around the world by the second week in September and stretched out to higher latitudes as the weeks went by, blanketing Europe by the end of November. In the whole of the last week of that month, Europeans were treated to vividly coloured skies and rose-tinted twilights that lingered long after sunset, while at the Montpellier Observatory in the south of France astronomers noted with astonishment a drop in their measurements of the Sun's direct radiation from 30 per cent above normal for the time of year to 20 per cent below normal. The direct solar energy measured by the observatory remained 10 per cent below normal for three years after the eruption, but this does not mean that a full 10 per cent of incoming solar heat was blocked by the dust in the stratosphere.

Only part of the incoming solar heat that seems to be lost is actually scattered back out into space by the dust. Some of the heat goes to warm the dust itself, in both the stratosphere and troposphere, and some is scattered sideways, so that it still reaches the ground even though it does not appear to be coming directly from the Sun's disc, which is what measurements of direct solar radiation actually record. If the Sun's output really reduced by 1 per cent, then the mean surface temperature would fall by about $1\frac{1}{2}$ or 2 degrees. Reid Bryson and his colleague Brian Goodman, writing in *Science* in 1980, estimated that during historical periods when the measured decrease in direct solar radiation was about 5 per cent the Earth's surface actually cooled not by nearly 10° C (sufficient to start a new Ice Age) but by less than one degree, because the decrease in direct radiation was almost, but not quite, matched by the increase in indirect radiation, scattered sideways by the dust.*

So, apart from the spectacular sunsets and other side effects, the influence of one or two Krakatoa-sized eruptions could certainly tip the balance between present-day climatic conditions and those of the Little Ice Age, and, of course, anything in between. When Wexler was writing about volcanoes and climate for *Scientific American* in 1952, he was more concerned with climatic changes in the first half of the twentieth century, the global warming out of nineteenth-century conditions that made the middle part of this century so unusual compared with the past millennium, and he remarked on 'the striking fact that since 1912 no major volcanic explosion has occurred in the Northern Hemisphere and during this period the winters have been growing steadily warmer'. He did not remark on the equally interesting fact that over the same period the Sun's activity had been increasing from one solar cycle to the next, and dismissed the suggestion of a link between solar activity and the weather by saying that the warming trend 'has marched on through three full sunspot cycles. The one conspicuous change has been that whereas during the 150 years before 1912 volcanoes erupted in one great explosion after another in the Northern Hemisphere, since 1912 they have been comparatively quiet.'

Like many enthusiastic pioneers, Wexler pressed his case a little too strongly. As we shall see, many of the features of climatic change in the past 300 years can be most neatly explained by a *combination* of the solar influence and the effects of volcanic dust on the atmosphere. But this only became clear after Hubert Lamb had painstakingly compiled his definitive Dust Veil Index.

Just how big an effect dust in the atmosphere has on climate depends on both the nature of the dust and just where in the atmosphere it is. Krakatoa, for example, was a less significant eruption than Tambora in climatic terms, even though it produced such

---

* These particular calculations by Bryson and Goodman relate not only to volcanic dust in the atmosphere but also to dust from human activities. This does not affect the argument about sideways scattering and the actual impact of Krakatoa on climate.

Table 3: The major volcanic eruptions from the time of Benjamin Franklin's observations to 1970, when Hubert Lamb published his Dust Veil Index

Year	Volcano	Situation		DVI
1783	Eldeyjar, off Iceland	63½°N	23°W	700
	Laki and Skaptar Jökull, Iceland	64°N	18°W	
1783	Asama, Japan	36½°N	138½°E	300
	Total veil, 1783:			1000
1786	Pavlov, Alaska	55½°N	162°W	150
1795	Pogrumnoy, Umanak Is., Aleutians	55°N	165°W	300
1796	Bogoslov, Aleutians	54°N	168°W	100
1799	Fuego, Guatemala	14¼°N	91°W	600
1803	Cotopaxi, Ecuador	1°S	78°W	1100 (?)
1807–10	Various, including Gunung			
	Merapi, Java	7½°N	110½°E	(?)
	and São Jorge, Azores	38½°N	28½°W	(?)
	Total veil, 1807–10:			1500 (?)
1811	Sabrina, Azores	38°N	25°W	200
1812	Soufrière, St Vincent	13½°N	61°W	300
1812	Awu, Great Sangihe, Celebes	3½°N	125½°E	300
1813	Vesuvius	41°N	14°E	100
1814	Mayon, Luzon	13½°N	123½°E	300
1815	Tambora, Sumbawa	8°S	118°E	3000
	Total veil, 1811–18:			4400
1821	Eyjafjallajökull, Iceland	63½°N	19½°W	100
1822	Galunggung, Java	7°S	108°E	500
1826	Kelud, Java	8°S	112½°E	300
1831	Giulia or Graham's Island	37°N	12–13°E	200
1831	Pichincha, Ecuador	0°S	78½°W	(?)
1831	Babuyan, Philippine Is.	19°N	122°E	300
1831	Barbados	13°N	60°W	(?)
	Total veil, 1831–33:			about 1000
1835	Coseguina, Nicaragua	13°N	87½°W	4000
1845	Hekla, Iceland	64°N	19½°W	250
1846	Armagora, S. Pacific	18°S	174°W	1000
1852	Gunung Api, Banda, Moluccas	4½°S	130°E	200
1856	Cotopaxi, Ecuador	1°S	78°W	700
1861	Makjan, Moluccas	½°N	127½°E	800
1875	Askja, Iceland	65°N	17°W	300
1878	Ghaie, New Ireland, Bismarck			
	Archipelago	4°S	152°E	possibly 1250
1883	Krakatoa	6°S	105½°E	1000
1888	Bandai San, Japan	38°N	140°E	250
1888	Ritter Is., Bismarck Archipelago	5½°S	148°E	250
	Total veil 1883–90:			about 1500
1902	Mount Pelée, Martinique	15°N	61°W	100
1902	Soufrière, St Vincent	13½°N	61°W	300

Table 3 – *cont.*

Year	Volcano	Situation		DVI
1902	Santa Maria, Guatemala	$14\frac{1}{2}°$N	$92°$W	600
	Total veil, 1902:			about 1000
1907	Shtyubelya Sopka/Ksudatch,			
	Kamchatka	$52°$N	$157\frac{1}{2}°$E	150
1912	Katmai, Alaska	$58°$N	$155°$W	150
1963	Mt Agung (Gunung Agung), Bali	$8\frac{1}{2}°$S	$115\frac{1}{2}°$E	800
1966	Awu, Great Sangihe, Celebes	$3\frac{1}{2}°$N	$125\frac{1}{2}°$E	150–200
1968	Fernandina, Galapagos	$\frac{1}{2}°$S	$92°$W	50–100
	Total veil, 1963–68:			about 1100
1970	Deception Is.	$63°$S	$60\frac{1}{2}°$W	(200)

The veiling effect on the atmosphere of the 1883 eruption of Krakatoa is arbitrarily set as 1,000, and other eruptions compared with this standard, using a variety of historical records. (Source: *Lamb*, Climate: Present, Past, and Future.)

a dramatic spell of dust.* Lamb divides eruptions that throw dust into the atmosphere into two categories, drawing a distinction between dust layers in the lower stratosphere, between about 20 and 27 km altitude, and those which reach higher, up to 50 km. Paradoxically, the *first* group is the more important in influencing climate, because such eruptions produce dense, long-lived dust veils. The rarer eruptions which throw dust up above 50 km are less important for the broad features of the changing climate, and for the average DVI, simply because so little dust ever gets that high. Today, too, many atmospheric scientists believe that it may not be the dust alone which blocks out heat from the Sun. Volcanoes also produce great quantities of gases, including oxides of sulphur which react with water to produce sulphuric acid and other droplets which may penetrate the stratosphere like a fine mist and obscure the Sun. But the effect is the same – bigger eruptions, those which rate a bigger DVI on Lamb's scale, are still the most important in climatic

---

* The earlier eruption also produced its spectacular sunsets, green Suns and blue Moons, as well as wrecking the harvests. One curious result is familiar today in the work of England's greatest artist, J. M. W. Turner, who produced some of his best works, full of colour and strange light, in the years after Tambora erupted. These works, so dramatically striking to us, were strongly influenced by the dramatic lighting he saw in the real world at the time.

terms, whether it is dust, or sulphuric acid, or a combination of the two, that does the job.

Lamb found that many of the coldest, wettest summers in Britain, North America and Japan have, like 1816 and 1784, been 'volcanic dust years'. He says that 'it may be that volcanic dust has played a part in all of the very worst summers of these centuries [the seventeenth to the twentieth]', and that 'it seems clear that there is also a tendency for cold winters in Europe after some volcanic eruptions, evidently those in low latitudes, that produce world-wide dust veils'. The key word here is *tendency*, for not even Lamb claims that the changing DVI alone is responsible for all the climatic changes of the past three or four centuries. Lamb's work, and other studies, show that volcanic dust is an important *contributor* to climatic change, but again, in Lamb's words, 'nevertheless, volcanic dust is not the only, and probably not the main, cause of climatic variation within the period surveyed'. This is borne out clearly by the survey by Schneider and Mass, which showed that a combination of the changing volcanic influence, measured by Lamb's DVI, and the changing solar influence, measured by sunspot number, best explains the pattern of climatic change since the Little Ice Age.

In this case, however, it may be that the pioneer has been a little too cautious in pushing forward the idea he favours, since, late in 1979, Alan Robock of the University of Maryland published in *Science* an analysis which suggests that the pattern of climatic changes studied by Schneider and Mass is indeed dominated by the volcanic changes, rather than the solar influence.

A broad perspective

All these interpretations seem to depend on which particular computer model of the way the atmosphere responds to different influences you favour, and none can yet be regarded as definitive. Robock's computer model includes an allowance for the changes in the albedo of the Earth caused by snow and ice fluctuations, an effect of solar variations which smooths out sunspot numbers over the whole 11-year cycle (but does not, unfortunately, take into account the evidence from Kondratyev and Nikolsky's work that very high sunspot numbers, like very low ones, are related to cooler conditions on Earth), volcanic dust effects with a DVI of 160 corresponding to a 0·5 per cent decrease in the direct radiation from the Sun, an allowance for the build-up of carbon dioxide in the atmosphere, and even a mixed-in random effect to make the whole thing more 'natural'. The numbers that come out of the machine match the actual record of temperature changes since 1600 best when the DVI is the only variable factor; the sunspot variations do not provide such good agreement. But that may be telling us more about this particular computer model than about Sun–weather relationships, since, as Robock says, 'the result does not rule out changes in the solar constant as causes of climatic changes, but, if there is a relation, these changes must ... [be related] ... to sunspots in some very complex way'. That complexity, involving changes in the transparency of the stratosphere, the link with galactic cosmic rays, and the reversal of the effect for sunspot numbers above about 80, is just what the Russian studies find.

If you are looking for complex relationships, some very curious ones have been uncovered by Mick Kelly and his colleagues at the University of East Anglia. Kelly, a former student of Lamb's, has looked for regular variations in both DVI and climate, and has found that both of them show a cycle 7 to 8 years long. It is unlikely that two such cycles could march in step simply by chance over the period studied (from 1725 to 1950), and this discovery alone supports the idea that changing volcanic activity influences the climate. The 7 to 8-year volcanic cycle must also be related to a similar known fluctuation in the Earth's rotation, in which the length of day first increases then decreases by a fraction of a second.

It is reasonable enough that wobbles of the Earth could trigger volcanic eruptions, and thereby affect the weather, even if the wobbles are small; however, writing with Lamb in *Nature* in 1976, Kelly commented on a much stranger discovery – a 180-year rhythm in volcanic activity resulting from changing tidal stresses acting on the Earth. The 180 (or 179) year cycle in weather records is familiar from studies like those of Hurd Willett, and many people have commented on the similar 179-year rhythm in solar activity. The changing tidal stresses on Earth are related to the long-term changes in the gravitational forces of the planets in the Solar System, acting on the Earth as they move in their separate orbits around the Sun. Some commentators have 'explained' the 179 (180) year solar rhythm as also due to the changing tidal influence of the planets, which repeats over a period exactly this long. And this raises the intriguing possibility that although both the Sun and the Earth are responding to the same 180-year rhythm, so that the Sun seems to be affecting the weather on Earth, in fact the sunspot variations and the volcanic/climatic changes on Earth may be produced independently!

That is enough to make anyone's head spin, whatever effect it has on the spin of the Earth, but it certainly adds to the interest of observers watching solar activity and the weather over the next couple of decades to test out the predictions made by Willett and outlined at the end of Chapter 5. From such esoteric complexities, though, it is something of a relief to come back down to Earth with an absolutely clear-cut picture of the relationship between the broad level of volcanic activity over the past 10,000 years and the temperature fluctuations of the Northern Hemisphere, a picture provided by our old friends the cores from the Greenland ice-sheet.

Willi Dansgaard's team is always looking for new ways to extract information from the Greenland ice-cores, and in 1980 a team from the Copenhagen laboratory, headed by C. U. Hammer, reported their successful attempts to detect traces of volcanic eruptions from the same cores which provide them with direct measurements of past temperatures from the changing proportions of oxygen isotopes in the ice. The volcanic traces were, if anything, easier to analyse than the oxygen isotope thermometer, once the idea had occurred to the team. Because volcanic eruptions eject huge quantities of acid gases into the atmosphere – where they may be as important as dust particles in cooling the Earth below – periods of great volcanic activity are also periods of acid rainfall, or acid snowfall in the case of the Greenland ice-sheet. The traces of acid in the snow which has fallen to build up the ice-sheet conduct electricity more easily than pure water, or ice, so they can be detected by measuring the electrical conductivity of melted ice samples, by passing an electric current through the solid ice, or by simply pulling a pair of electrodes with a voltage difference of 1,250 V along the clean-cut surface of an ice-core and recording the changing electric current. This fast, effective method has the advantage of leaving the core intact for other studies, such as isotope measurements to determine the pattern of past temperatures, and the result is an unambiguous measure from the same ice-core both of past global volcanism and of past temperatures.

Comparison of the measurements from the top layers of the ice with the historical record of volcanic activity shows that the acid test is a very good indicator of volcanic activity in latitudes north of 20° S.

These are, in any case, the ones important for the climate of the Northern Hemisphere, where most of the continents, and the human population, of the Earth reside. The acidity record shows that persistent volcanic activity north of 20° S has been an important, but, as the Danes point out, not necessarily the only, cause of 'climatic fluctuations of up to several hundred years duration at mid and high latitudes' of the Northern Hemisphere for the past 10,000 years. The acidity changes match up not only with temperature changes from the Greenland cores, but with changes estimated from tree-rings, and with the longest available record of temperatures and estimated temperatures from central England.

The measurements also tie in some anecdotal historical material to the volcanic record, with a high acidity around 50 BC suggesting one of the largest eruptions in the Northern Hemisphere since the latest full Ice Age. Pliny the Elder tells us that in the year of Caesar's death (44 BC) the Sun shone feebly 'through nearly a year, when the dictator Caesar was killed', and Virgil states that 'when Caesar died, the Sun felt pity for Rome, as it covered its beaming face by darkness, and the impious generation feared an eternal night'. This curious pre-echo of Ben Franklin's remarks in 1784 suggests a major volcanic eruption somewhere in the world, an eruption now confirmed by the acid record in the ice-cores. But if eruptions down the centuries can go at least some of the way towards explaining climatic fluctuations on the scale of the Little Ice Age, and if individual eruptions, or clusters of eruptions, in 44 BC, 1783 and 1815 have had such dramatic impacts on the environment and mankind, how have we managed to survive the equally dramatic eruptions of Mount St Helens virtually unscathed?

St Helens: a lucky escape

The summer of 1981 missed being like that of 1816 for one main reason. Unusually, the blast from Mount St Helens was directed sideways, not upwards, as one wall of the mountain collapsed and released the accumulated pressure inside with explosive violence. The explosion on 18 May 1980, which has been estimated as equivalent to between 10 and 50 Mt of TNT, devastated an area of some 400 square km, and blast effects extended 22 km to the north-west and 19 km to the north. But little of the $2\frac{1}{2}$ cubic km of rock pulverized into ash by the force, equivalent to more than 500 times the explosive power of the Hiroshima atomic bomb, was directed upwards into the stratosphere. No globe-encircling high-altitude dust cloud developed, and although there were measurable effects of the eruption on the stratosphere, the loss of heat reaching the ground from the Sun, and the effects on weather during 1981, were minimal.

Even if the dust from St Helens had been blasted vertically into the atmosphere, the effects would probably not have been as severe as those following the eruption of Tambora in 1815, because the historical records show, and modern computer modelling on high-speed computers confirms, that dust injected into the stratosphere in the tropical regions spreads more and has a bigger effect than dust injected by volcanoes at high latitudes like St Helens. But overall, in a series of eruptions during 1980, St Helens blasted out a mass of dust estimated by Professor Lamb as comfortably more than the dust erupted from Krakatoa in 1883. So the latitude of the volcano could not on its own have saved us from bad weather and ruined crops in 1981, had it not been for the lucky accident that the dust blasted sideways, so that it choked the troposphere downwind from St Helens, rather than the stratosphere above, and was soon washed out of the atmosphere by rain.

The best way to get a handle on what we missed is to look again at the events of 1816. These have been summarized by Henry and Elizabeth Stommel, a husband-and-wife team at the Woods Hole Oceanographic Institution. The Stommels plan to publish a book about 'the year without a summer' shortly, but meanwhile they have written about their findings in a *Scientific American* article which provides the best succinct account of the effects of Tambora on the weather of North America and Europe.

Hubert Lamb rates the eruption of Tambora in 1815 as one of the two greatest single producers of stratospheric dust between 1600 and the present day. The height of Mt Tambora was reduced by nearly 1,400 metres, and the eruption ejected about 100 cubic km of debris, not all of it as fine ash reaching the stratosphere. There are no good records of how the dust spread around the world, but the pattern must have been similar to the spread of dust from Krakatoa shown on page 136 (Figure 6.1). Mariners reported finding large islands of floating pumice from the eruption up to four years after Tambora exploded. As the dust spread to higher latitudes, the temperate zone experienced not a slow, steady cooling but waves of bitter cold which first killed the crops in the fields and then ruined the repeated efforts of farmers to salvage something from the wreckage.

In New England a cold and backward spring of 1816 caused no more than the usual grumbles among farmers, and a good beginning to June seemed at first to have made up for the frosts of May. But a wave of cold moved eastwards into New England from 6 June onwards, lasting for five days and leaving several inches of snow on the ground. Today we can identify the wave of cold with a tongue of the expanded circumpolar vortex licking down from the frozen North; in 1816 people could only wonder at the strange turn of events, as the *North Star* of Danville, Vermont, reported:

> Wednesday last [5 June] was perhaps as warm and sultry a day as we have had since September – At night heat lightning was observed, but on Thursday morning the change of weather was so great that a fire was not only comfortable, but actually necessary. The wind during the whole day was as piercing and cold as it usually is the first of November and April. Snow and hail began to fall about ten o'clock, A.M., and the storm continued until evening, accompanied with a brisk wind, which rendered the habiliments of winter necessary for the comfort of those exposed to it . . . probably no one living in the country ever witnessed such weather, especially of so long continuance.

But it wasn't to be long before they experienced its like again. No sooner had the baffled farmers started to repair the damage from the week of unseasonable cold than, in early July, a second cold wave swept the region. This was less severe than the killing frosts and snows of June, but more remarkable in that it happened at all even later in the summer. Ice and frost killed crops between 5 and 9 July, but summer weather returned on the 12th and it remained pleasantly warm until 20 August. By then the harvest of what crops remained after the ravages of the two cold waves was about to begin. But now came the worst of the unseasonable summer cold of the whole year. Frost killed crops in New Hampshire and Maine and was a problem even further south in Massachusetts; the mountains of Vermont were covered with snow, and the corn crops in the plains were ruined. Hardly surprisingly, conditions were worse further north in Canada, where even wheat, which had survived reasonably well in the US, perished.

In a Europe devastated by the last stages of the Napoleonic Wars, which had ended in 1815 with the Battle of Waterloo and Napoleon's exile to St Helena, agriculture and the economy were wrecked by the weather of 1816. The French in particular had no food reserves to tide them over even an ordinary bad summer, and many regions were affected by famine and rioting. Over the whole period from 1801 to 1912, the highest wheat prices in France were recorded in 1817, twice as high as the long-term average. And as Europe turned to America to buy grain, prices there rose too. The failure of the corn crop from Pennsylvania northwards, plus the demand from Europe, saw the price of wheat soar from an average $1.30 per bushel in the first decade of the nineteenth century to $2.45 in 1817 before falling back to a steady level of just over a dollar a bushel for the next three decades. With flour being shipped to England and mainland Europe at inflated prices, some dealers made a killing while many ordinary folk went without. Food shortages and severe weather combined to encourage many people to make the trek westwards into the virgin lands; to take one example, the number of prospective pioneers leaving Vermont in 1816–17 was almost twice that in other years of the decade, while over in Ohio the Zanesville *Messenger* noted in its edition of 31 October 1816 that 'the number of Emigrants from the eastward the present season, far exceeds what has ever before been witnessed'.

All this thanks to the eruption of one volcano on the other side of the world. But perhaps the strangest aspect of the whole story is that no one seems to have made the connection between the two events. Some nineteenth-century scientists blamed the bad weather on sunspots, or the lack of them; others attributed the events of 1816 to the spread of icebergs south from the Arctic. One bizarre idea did link the year without a summer with Benjamin Franklin, but not with his account of atmospheric dust in 1783–4. This curious theory held that the Earth was kept warm not only by the heat of the Sun but by 'electric heat' escaping from its interior. The introduction of lightning rods, invented by Franklin, had, so the argument ran, disrupted the natural flow of this internal electricity and thereby disrupted the weather. Had Franklin still been around, he would surely have set people right; as it was, in many newspapers reports of the bad weather appeared alongside reports of large floating islands of volcanic ash seen in the Pacific. But only in the second half of the twentieth century did scientists connect the two as cause and effect.

No individual eruption of St Helens could hold a candle to the single, massive explosion of Tambora, or even to the great eruption of Krakatoa. But St Helens did remain active throughout 1980, with six main phases of explosive eruption, the first of which was by a long way the biggest and most spectacular. The initial eruption ripped 400 metres off the top of the mountain, putting it in the same league as Tambora but about one third as impressive. The explosion was heard 320 km away and produced a plume of ash rising 14,500 metres into the sky. Even 800 km away from the mountain acidic ash, the fall-out from the explosion, piled up to a depth of $1\frac{1}{2}$ cm, and by 20 May, 2,000 km away in Manitoba and Ontario, the ashfall reduced visibility to 5 km. To put it in more familiar everyday terms, the ash plume contained 1·3 *billion* cubic yards of material – which is a great deal of ash, as anyone who has ever purchased a yard of sand from a builder will appreciate. The total cost of the damage was more than $\$1\frac{1}{2}$ billion, with crops, orchards and timber devastated, while 23 ships were stranded by the build-up of silt in Portland harbour. But all this ash, which caused so much inconvenience to so many people, and provided spectacular TV film for the rest of us, was ash which *failed* to get into the stratosphere. The very fact that so many people in the northern US and Canada were directly affected by the eruption, thanks to its sideways blast, meant that the rest of the world did not have to suffer a summer like that of 1816. Even so, the stratosphere did not get off scot-free.

Cameras carried on balloons launched into the stratosphere by a joint Belgian/French team on 7 May and 5 June 1980 clearly show the obscuring effect of material from the Mount St Helens eruption. In May, before the pollution reached Europe, the pictures taken show clear skies and good visibility at 15 km altitude, producing a breath-taking panorama across the tops of the clouds in the troposphere below. In June, pictures taken under identical circumstances show only an obscuring haze. Other instruments mounted on the balloons show that this haze was made up of millions of tiny particles, called aerosols.* The aerosol layers ranged in thickness from 100 metres to several kilometres, but they dominated the region from a few kilo-metres altitude up to about 15 km, with a very sharp cut-off above. These layers must have absorbed solar heat that might otherwise have reached the ground, but they could not produce a dramatic

---

* The familiar 'aerosol sprays' are so called because they produce a fine mist of similarly sized particles – fine enough to be suspended in the air, hence the name.

global cooling because most of the energy was still being absorbed in the troposphere, the weather layer of the atmosphere. The evidence is that material from Mount St Helens – dust, sulphuric acid droplets or whatever – simply did not reach the altitudes around 25 km which Hubert Lamb has shown to be all-important if the dust veil is to influence the climate.

Putting together all the best observations of the dust from St Helens in the months following the eruption, and feeding the numbers into his computer model of the response of the atmosphere to dust veils, Alan Robock was able to calculate just what effect the eruption had on our weather. The climate model he uses takes account of the latitude of the eruption and the slow spread of the dust veil around the world, as well as the thickness of the layers and their measured effect on incoming solar radiation. And it came up with the calculation that the biggest effect on global climate was a mere $0.1°$ C cooling in the North Polar regions for the month of January 1982. Since winter temperatures over the North Pole can easily vary from year to year by $2\frac{1}{2}$ degrees due to all the other factors affecting the weather, what Robock's calculations are really telling us is that the effect of Mount St Helens on our weather has been too small to measure. Other twentieth-century volcanoes, however, have played a much bigger role in determining weather and climate.

The twentieth century and the human volcano

After the end of the nineteenth century the world as a whole (measurements are actually for the Northern Hemisphere, but they are thought to indicate a global trend) first warmed up slightly by about half a degree centigrade, and then cooled down after about 1940. By the 1970s the cooling had reached $0.5°$ C from the peak of the early 1940s. To some extent, as we have seen, this pattern may be related to changes in solar activity. But another factor which must have played some part – the puzzle is how big a part – is the changing pattern of volcanic activity. Lamb's calculations of the Dust Veil Index (see pages 139–40) show how the middle part of the twentieth century was relatively free from the cooling effects of volcanic dust. After several eruptions in the late nineteenth and early twentieth centuries, including Krakatoa, there were relatively few volcanic outbursts for several decades. It could be, some climatologists argue, that the warming of the world up to the 1940s corresponds to the time it took for volcanic dust to clear out of the stratosphere. In the second half of the century volcanic activity has certainly been on the increase, with notable eruptions of Mount Spurr in Alaska in 1953, Mount Bezymiannyi in Kamchatka in 1956, Mount Agung on Bali in 1963, and Mount St Helens in 1980. Not all of these eruptions are on Lamb's list of important contributors to the Dust Veil, but they show, along with others, that the Earth has been more seismically active lately following a period of unusual quiet.

If dust is to be invoked as the prime cause of climatic changes since the nineteenth century, however, there is one remaining puzzle to be resolved. Why did the globe begin to cool in the 1940s, a decade before the upturn in volcanic activity? Reid Bryson of the University of Wisconsin – Madison – thinks he has the answer.

Professor Bryson is a confirmed supporter of the theory that fiery volcanoes cause icy conditions on Earth. In *Climate of Hunger* (page 148) he summarizes the theory as follows:

> Dust in the atmosphere tends to cool the high latitudes more than it does the tropical regions, no matter where it enters the atmosphere. Some dust is carried poleward, over a period of weeks and months, by the high altitude flow of air from the tropics. More important, even if dust were distributed evenly throughout the atmosphere, the poleward regions would be more shaded by it than the tropics. Sunlight takes a nearly vertical path through the atmosphere in the tropics, but away from them comes in at an angle, and therefore has a longer path through the atmosphere. If the atmosphere is dusty, the sunlight has a longer path through the dust – and is diminished all along the way.

And Bryson explains the size of the cooling since 1940, and the reason for its onset before the upturn in volcanic activity, as a result of human activity. Pollution produced by people is also putting dust into the air, including windblown soil from farmland, pollution from factory chimneys and car and aeroplane exhausts, dust from barren lands ruined by bad farming practices, smoke from slash-and-burn agriculture, and so on. All this, says Bryson, is equivalent to a continuously erupting 'human volcano', adding its burden to the atmosphere. The amount of dust suspended in the atmosphere at any one time as a result of the human volcano is estimated by Bryson at 15 million tonnes, and he believes that this is the dominant influence on climate today, producing a continuing global cooling.

The human volcano idea is not without its critics. A great deal depends on just how big the dust particles produced by human activity are, and just where in the atmosphere they are concentrated. As Bryson himself points out, large dust particles can trap heat from the ground that would otherwise be radiated out into space, and can thereby warm the world. It is the smaller particles – the aerosols – that tend to let heat from the ground out into space, while stopping heat from the Sun from reaching the ground. The American climatologist Murray Mitchell estimates that human activities produce, directly or indirectly, about 30 per cent of the total atmospheric loading of particles less than 5 microns (five millionths of a metre) across, and since the total load is about 40 million tonnes this fits in reasonably well with Bryson's estimate of an anthropogenic influence amounting to 15 million tonnes. The key question, then, is whether these particles lie in regions of the atmosphere over light- or dark-coloured regions of the globe beneath.

Put simply, 'grey' dust overlaying a white surface will absorb more incoming solar heat than the white surface – a snowfield perhaps – would itself, so the world warms; the same grey dust overlaying a dark surface – a ploughed field perhaps – absorbs less heat than the dark surface would, and the ground cools. Ruth Reck of the General Motors Research Laboratories in Warren, Michigan, is one of the researchers who have put numbers into this simple-minded picture, and she comes out with the conclusion that for underlying surface albedos greater than 0·6 (that is, 60 per cent of incident radiation reflected) the effect is always a warming, and for underlying surface albedos less than 0·5 the effect is always a cooling, regardless of the height of the aerosol layer above the clouds. The jury is still out on the human volcano hypothesis, but expert opinion today seems to be that, while anthropogenic dust may play a part in cooling the globe, it is not yet as important as the dust veil from volcanic eruptions. It does, however, seem unlikely that the human volcano effect is providing a net warming influence on the globe; whatever effect it does have must indeed be a cooling one, even if it is not as large as Bryson and his colleagues calculate.

So far, and especially with Mount St Helens proving an exception to the volcanic norm in so many ways, the Mount Agung volcanic eruption of 1963 provides the best-documented evidence of how dust from volcanoes affects the transmission of radiation from the Sun through the atmosphere of the Earth. The one thing Mount St Helens did make clear is the importance of sulphuric acid droplets, not just dust, in the stratospheric aerosol layers. To put things in perspective, the amount of aerosol formed by the St Helens eruption (made up of particles from 0·1 to 0·6 microns in diameter) was roughly half a million tonnes. This represents a doubling of the usual 'load' of the stratosphere – but Agung is estimated to have contributed to the formation of 20 times as much stratospheric aerosol, from an eruption roughly the same size as the 18 May eruption of St Helens but directed upwards into the stratosphere.*

After the Mount Agung eruption the temperature of the troposphere between 30° N and 30° S fell by almost half a degree centigrade by late 1964, and then recovered to about its pre-eruption average by the end of 1966. In the late 1970s climate-modellers were pleased that their computer calculations of the weather machine's response to the volcanic aerosol burden could produce almost exactly the required pattern of behaviour – being able to mimic the real response of the atmosphere to a real volcanic eruption gives them confidence when using the same computer models to predict how the world would respond to hypothetical disturbances such as the human volcano or a build-up of carbon dioxide. But late in 1979 James Coakley of the US National Center for Atmospheric Research reported the latest, and best, calculations of this kind. He used Australian comparisons of direct observations of the heat from the Sun's disc with observations of the heat radiation coming from the rest of the sky when the Sun's disc was obscured, and he deduced that the Agung

---

* The massive eruption of the volcano El Chicón in Mexico in March 1982 is now providing climatologists and atmospheric scientists with their best-ever guide to how such events affect the Earth. As yet (October 1982), it is too early for the experts to have assessed the impact of the eruption on temperatures in the Northern Hemisphere, but the expectation is that it will tend to produce cooler winters and wetter summers over the next few seasons. The key questions still outstanding are 'how cool' and 'how wet'. By the time you read this we may know the answer.

eruption 'caused a lower-atmospheric cooling more than double that predicted by theory'.

The numbers are familiar, for we saw in Chapter 5 how Kondratyev and Nikolsky explain the cooling of the mid 1960s in terms of the effect of nitrogen oxides from nuclear bomb tests on the stratosphere. They claim that the overall cooling is roughly twice what can be explained by the nuclear debris alone, and that therefore the Agung eruption must have contributed an equal amount of cooling to the disturbances caused by the bomb tests. In other words, the actual cooling is twice the effect of Agung alone – otherwise, if you were to try to explain the effect by Agung alone, you would have to invoke an influence of the volcano on climate of twice its actual influence. This fits rather neatly with Coakley's calculations: leaving Agung's contribution as predicted by theory, the remaining influence found by the Australian observations can be explained by the nuclear-bomb effect reported by the Russians. And, leaving aside the anthropogenic influences, the human volcano and the bomb tests themselves, this strengthens the case that on a time-scale of decades and centuries small climatic changes are best explained by a combination of volcanic dust and solar influences on the stratosphere. That may leave the theorists happy, but, as expected, it doesn't leave us with any more encouraging forecast of the natural trend of climate in the decades ahead.

With normal volcanic service apparently restored after the quiet of the decades ending with the 1940s, yet another natural influence on climate is now acting to produce a cooling. With the possible solar influences, plus the probable contribution of the human volcano, it is easy to see why many climatologists have recently been concerned about the prospect of a continued global cooling, and why, ironically, it is by no means obvious that any warming influence due to human activities – the carbon dioxide greenhouse effect – should be regarded as entirely a bad thing, in the short term at least. Before we look in detail at the global greenhouse, however, one more natural influence on climate deserves a mention. It may not operate entirely on a time-scale important over a human lifetime, and the evidence may not yet be as clear-cut as the evidence in favour of the Milankovitch Model of climatic change. But the remarkable story of the discovery of a relationship between the Earth's changing magnetic field and the changing climate is certainly worth telling in its own right.

7
The Magnetic Link

A less determined man than Goesta Wollin might have given up his
studies of the links between geomagnetism and climate when his first
publications on the subject, in the early 1970s, met with a decidedly
cool response from most of the scientific community. Geophysicists
poured scorn on his idea that a correlation between climatic change
and changes in the Earth's magnetic field could be more than a mere
coincidence or statistical fluke. But the detractors should have realized that a man who, although a Swedish citizen, chose to join the
US Army to fight against Adolf Hitler, and whose first experience
of parachute jumping was to be dropped into Normandy the night
before D-Day as a member of the Intelligence Service of the 82nd
Airborne Division, was unlikely to be put off by a little academic
sniping.

Wollin's story – and that of the link between the Earth's magnetism and climate – deserves telling in its own right. But this is very
much an area of active research in science today, and it is impossible
to dot the i's and cross the t's with the same sense of satisfaction as
when telling the story of, say, the Milankovitch rhythms of climate.
I tell it here as much to give a flavour of the way progress is made
at the cutting edge of research, especially in such a multidisciplinary
area as the study of climate, as through any feeling that the ideas are
fully tested. Some people might argue that scientific debate should
be kept within the halls of academe – and the pages of learned
journals – until all the details have been settled. But if popular
accounts deal only with the cut and dried, proved and tested ideas
they miss out on the real excitement of science: the new discoveries
and the new ideas. The links between magnetism and climate are
certainly established well enough for this particular exciting story to
be told; what is not established to anyone's satisfaction is how the

links work. Even the progress so far, however, has represented a saga of modern scientific endeavour.

Looking back now at the controversy he stirred, Wollin clearly relishes the tale; but it has taken ten years of intensive work and sometimes bitter academic wrangling – once bringing him close to a nervous breakdown – to establish the reality of the links between geomagnetism and climate. Wollin stresses the importance of the distinction between the fact – now established beyond reasonable doubt – that magnetic and climatic changes are correlated, and the hypotheses, as yet still somewhat tentative, put forward to explain the correlations. Like other suggestions, the latest hypothetical 'mechanisms' may be proved wrong, even though they look good today. But if they are proved wrong, that only means we have yet to find the real mechanism; it does not mean that the correlations are not real.

Although suggestions of a link between geomagnetism and climate go back at least thirty years, the modern version of the story began with a serendipitous discovery by Wollin in 1970. He had been working for two decades at Columbia University in New York with David Ericson ('Eric') on, among other things, a study of long-period climatic changes during the Pleistocene, the most recent Ice Age epoch. In 1970 they turned their attention for the first time to short-period climatic changes, which on this time-scale means those which have occurred during the 11,000 years or so since the end of the latest full Ice Age. Their technique depended on analysis of the abundance of remains of creatures called foraminifera in the sediments of lakes or ocean bottoms. Foraminifera are tiny marine animals that float near the surface of the water and leave their microscopic shells behind in the sediments when they die. Different species prefer different water temperatures, so by the painstaking process of counting shells under a microscope the researchers can develop a reliable guide to the water temperature at the time when a particular community of creatures was flourishing. And, of course, isotope analysis of the oxygen in the chalky shells of the sediments also gives a handle on temperature – this was the technique used by the Lamont-Doherty team which provided the crucial, convincing evidence for the presence of the Milankovitch rhythms in the climatic patterns of the Earth's recent past. It is hardly surprising that Wollin and his colleagues, using the same kinds of skills to tackle similar problems, are also based at the Lamont-Doherty Geological Observatory, which is a part of Columbia University. The surprise is that Wollin and Ericson's first attempts at unravelling recent climatic changes from the sediments should have come to fruition just as geophysicists were unravelling details of the Earth's changing magnetic field over the same time span.

In April 1970, just when Eric had finished plotting his set of temperature curves for the past 11,000 years, a team headed by V. Bucha published in *Science* a curve showing changes in the intensity of the Earth's magnetic field over the past 9,000 years. Wollin describes how he returned from a trip to the northernmost ski area of Vermont, 650 km away, and thumbed through the latest issue of *Science* to wind down after the drive. The curve in Bucha's paper looked familiar and, unable to sleep, he took the copy of *Science* over to the Old Core Laboratory where he worked with Eric. There he compared the curve in *Science* with the six graphs indicating the changing climate which Eric had plotted using data from different sediment cores. But in his exhausted state, as he now ruefully admits, he found no agreement between them and the magnetic curve after all. At 7 A.M. he gave up looking at wiggly lines on sheets of graph paper and went to sleep on the lab floor; at 9 Eric arrived for work and woke him. Wollin started to talk about his trip to Vermont but stopped, suddenly aware of why the curve in the *Science* paper had looked so familiar – it was a mirror image of Eric's climate curves!

As many scientists have found, often with far more complex problems than this, 'sleeping on it' really does help with finding a solution. While the conscious mind is asleep, the unconscious mind seems to keep working on the problem, ready to pop up with the answer almost on request. But deceptive coincidences are the bane of scientists' lives, and many a beautiful pair of curves which seem to match perfectly over a particular stretch of data later turn out to have nothing to do with one another at all. In this case the agreement was striking, once the magnetic curve had been reversed. But Wollin and Ericson needed far more data to test the implied relationship – that when the Earth's magnetic field is *weaker* the climate is *warmer*. It would have been all too easy, especially in view of his exhaustion at the time, for Wollin to dismiss the apparent correlation as a coincidence not worth following up, but instead he and Ericson went ahead with further investigations, kicking off what has proved to be a whole new ballgame in climatic studies.

Links in the chain

Several different techniques came together in laying the foundations for the new work, partly thanks to the fact that geologists and geophysicists with a wide range of interests all work together at Lamont. By 1970 the geophysicists were well aware that the polarity of the Earth's magnetic field reverses from time to time, with North and South magnetic poles changing sign. These changes in polarity are revealed by the fossil magnetism in rocks laid down long ago, because sediments which contain magnetic rocks line up with the Earth's magnetic field while they are being laid down, so that once they have set solid their remaining magnetism provides a permanent record of the Earth's magnetic field during the time they were being deposited. Different sediments of different ages combine to give a picture of the Earth's changing magnetism.

Studies of fossil magnetism in the rocks of the sea-bed played a crucial part in establishing the reality of sea-floor spreading and plate tectonics – continental drift – in the early 1960s. Wollin and his colleagues were more interested in the changes in strength and polarity of the magnetic field revealed by the record in the rocks than in the sea-floor spreading process, but, thanks to the work done in connection with developing the ideas of plate tectonics, they had a wealth of material gathered for sea-floor studies to use in their work.

Working with Ericson, Wollin had developed, using the magnetic reversals as a calendar, a detailed time-scale against which geological events of the past two million years – the Pleistocene – could be compared. The pattern of magnetic reversals shows up in good samples of magnetic rock as unambiguously as a fingerprint or a tree-ring sequence, and once such a sequence has been dated using one sequence of well-understood rocks from a site on land, then the presence of the same pattern in the magnetism of, say, sediments from the sea-bed shows exactly when different layers in those sediments were being laid down.

By 1970 this Pleistocene magnetic calendar was well established. But the hint of a relationship between magnetism and climate came from a much shorter span of the recent past, just a few thousand years. To check out the implications, Wollin and Eric needed the equivalent of a fine-toothed comb, and the one they found was Willi Dansgaard's isotope record of changing temperatures from the Greenland ice-cap. The core provided a check of the relationship over the past 11,000 years; but when Wollin and Ericson tried to bridge the gap between that sequence and their studies of the past few million years they ran into their first brick wall. According to the best geophysical evidence available in 1970, it seemed that there had been no geomagnetic reversals during the past 700,000 years, a time of considerable climatic fluctuations (as we have seen!) which, if the magnetic theory were to hold water at all, must also have been a time of magnetic variability.

It wasn't until August 1970 that a chance meeting with William Ryan, a research student, led to a conversation in which Wollin learned that the basis of Ryan's Ph.D. thesis was to be his study of magnetic reversals during the past 470,000 years – a period in which he had found evidence, from two sediment cores, of five short-lived magnetic reversals. Moreover, one of the cores he had used, which came from the Caribbean, was one that Eric had already used for climatic analysis. Ryan's plot of magnetic fluctuations and Eric's plot of climate fluctuations, deduced from the same core, matched exactly; Ryan and his colleague John Foster promptly joined the team.

Over the next few months the four of them worked intensively on the project, with the Caribbean core as what Wollin calls their 'star performer'. It provided them with correlated variations in four different parameters: the intensity of the Earth's magnetic field, its inclination ('angle of dip'), temperature deduced from oxygen isotopes, and temperature deduced from foraminifera shells. The curves from this core study also matched up well with data from similar cores from the Pacific and Mediterranean, extending the net around the world (Figure 7.1).

It was with what seemed justifiable pride that the team despatched a paper reporting its findings to *Science* in March 1971. Its surprise when it was asked to revise the paper to meet a referee's objections was matched only by its consternation when the revised paper was rejected. But these things happen in science, and Wollin sent the original version of the paper off to another journal, *Earth and Planetary Science Letters*, where it was promptly accepted unchanged and in due course published. Meanwhile, however, Wollin had heard again from *Science*, in circumstances he recalls today with undisguised glee.

At the same time as the original paper had first been sent to *Science*, Wollin had sent a copy to the eminent climatologist Cesare Emiliani in Miami. Emiliani had done the pathfinding work in the 1950s

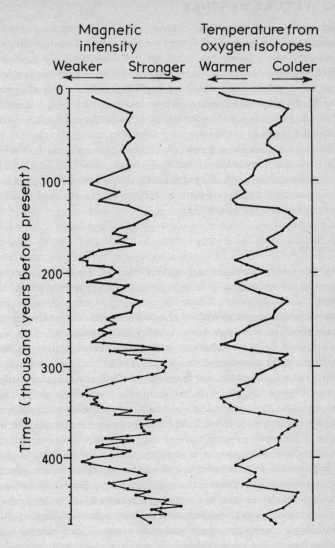

Figure 7.1. A single core drilled from the floor of the Caribbean provided the Lamont team with a direct comparison of changes in the Earth's magnetic field and temperature changes over almost half a million years. The correlation is unambiguously clear.

which first used studies of foraminifera in sediments as climatic indicators – he virtually invented the technique. The day after the Lamont team's paper had been formally accepted by *Earth and Planetary Science Letters*, Emiliani telephoned Wollin to report a conversation he had just had with John Ringle, then editor of *Science*. Emiliani had called Ringle to ask when the paper by Wollin, Ericson, Ryan and Foster would be appearing in print. Astonished to be told that it had been rejected (twice!), Emiliani informed Ringle that this was a highly important piece of work and *Science* should publish it. Sure enough, a letter from Ringle soon reached Wollin, inviting him to resubmit the manuscript. 'I must say,' Wollin reminisces, 'it was a pleasure to tell Ringle that my colleagues and I were not interested in resubmitting the paper!'

Growing controversy

But this was far from the end of their troubles. The *Science* incident highlighted what was to be a recurring theme: geophysicists dismissed the magnetism/climate relationships as coincidence, while many climatologists welcomed the evidence that the changing climate could be related to other geophysical phenomena. The controversy was heightened because the mechanism originally proposed for the link did not stand up to scrutiny. Wollin and his team argued that, perhaps, when the Earth's magnetic field is weaker and more cosmic rays can therefore penetrate into the lower atmosphere, they act to warm the world. Not only has no one found a satisfactory physical reason why this should be so, but more recent studies like those of Kondratyev and Nikolsky suggest at first sight that if anything such an effect ought to work the other way round – a weak magnetic field with reduced shielding against cosmic rays might be expected to produce similar effects to very intense bursts of solar activity, that is, a *cooling* of the globe. Theorists trying to unravel all these relationships are dealing with a very tangled web indeed. But the difficulty of explaining the observed facts does not falsify the facts themselves.

In the face of sometimes hostile opposition from geophysicists who specialized in the study of the Earth's magnetism, Wollin's team extended its correlations between magnetism and climate to cover in detail the entire 2 million years of the Pleistocene record. At the same time, many climatologists, including some at Lamont, were

deeply involved in the studies which were to establish beyond reasonable doubt the importance of the Milankovitch cycles to the
climate of the Pleistocene. And on the other side of the Atlantic
another scientific loner was coming up with intriguing evidence of
the reality of links between geomagnetism and climate operating on
much shorter time-scales.

Joe King is a researcher at the Appleton Laboratory in Britain.
This is primarily a centre for studying the ionosphere, and the way
radio waves propagate around the world, but King is one of those
rare scientific birds whose track record has proved so good that even
within the confines of a government laboratory, as a member of the
Scientific Civil Service, he has what is called a 'special merit' ap-
pointment giving him considerable freedom to follow his nose in
seeking out interesting and potentially important topics of research.*
The direction he followed in the early and middle 1970s took him
from an interest in the effects of solar activity on the weather, perhaps
involving the Earth's magnetic field (a topic not that far removed
from the Appleton Lab's main interests), to the puzzle of relationships between the shape and strength of the magnetic field and
weather and climate on Earth. The arrival of cosmic rays from the
Sun and the Galaxy disturbs both radio propagation and the weather, as we have seen. But on the way into the atmosphere they are
guided by the Earth's magnetic field, so it took no great leap of
intuition to look at how the magnetic field patterns of the Earth
compare with the circulation patterns of the atmosphere. The surprise is that they match up almost perfectly.

The magnetic field of the Earth is not as simple as the field of a
bar magnet; in particular, in the Northern Hemisphere today there
is a kind of double, or double-lobed, magnetic 'North Pole'. One of
these lobes of the double pole almost exactly coincides with the
centre of low pressure which dominates the circulation pattern over
the North Atlantic, and that may well be why, as many studies show,
these latitudes are particularly sensitive to changes in the Sun's
activity. The really curious thing, however, is that this atmospheric
circulation pattern, which can be mapped by plotting as contour
lines the changing altitude around the globe of the point in the

* Back in the 1960s, before he founded the Climatic Research Unit in
East Anglia, Hubert Lamb held a similar 'special merit' post at the UK
Meteorological Office.

atmosphere where the pressure is 500 millibars, also has a double-lobed structure which almost, but not quite, overlays the magnetic field structure. (See Figure 3.5, p. 80.)

King decided that this could not be coincidence, but must represent cause and effect, with the Earth's magnetic field somehow forcing the circulation pattern into a similar shape. He coined the name 'magnetometeorology' (perhaps a little prematurely) for the study of this and related phenomena, and he developed the idea by pointing out that both the magnetic field and the circulation pattern seem to be drifting from east to west around the globe in the present geological epoch. The westward drift of the magnetic field is already well known; King identified the westward drift of climate in particular from the way successive waves of cold (including, perhaps, the Little Ice Age) seem to have developed in China and Japan and moved across Russia into Europe during historical times.

Nothing much seems to have come of King's invention of the subject of magnetometeorology. Meteorologists don't seem to trust the magnetic side, while scientists interested in the magnetosphere don't seem to be very interested in the weather. In a review in *Nature* in January 1974, Hubert Lamb did comment that 'no such strong hints of association as King points to should be left uninvestigated', but left uninvestigated they by and large have been.

There, for a time, the case rested. But all evidence of links between geomagnetism and climate adds strength to studies such as those by the Lamont team. In June 1974, also in a review in *Nature*, Peter Smith of the Open University summarized some of what was by then an impressive weight of evidence gathered by Wollin and his colleagues, including their extension of the correlations, using different deep-sea cores, back beyond one million years into the past. 'Other workers are also finding relationships between magnetism and climate,' said Smith; 'it is by no means clear that these will be . . . easily refuted.'

Support from a geophysicist like Smith was particularly welcome for Wollin, Eric and the others. By the mid 1970s they were working towards an understanding of the way weather and geomagnetism interact on a time-scale of years and decades, rather than thousands and millions of years, with the Holy Grail of some form of improved long-range forecasting ability seeming, tantalizingly, almost within their grasp. But it remained just beyond reach for more than half a decade, during which time Wollin, Ryan and

Ericson came up, almost as a sideline, with a neat, if speculative, idea linking some of their findings with the growing success of the Milankovitch Model.

Orbital and magnetic changes

In the mid 1970s the 'worst' aspect of the Milankovitch Model seemed to be that it was 'too good to be true' – that the 100,000-year rhythm found in the climatic record was too strong to be accounted for by the orbital changes alone. As we saw in Chapter 4, this argument has been weakened by more recent analyses of the way ocean currents and sea-ice feed back energy into the Milankovitch long-term rhythm. Wollin and his colleagues offer another explanation, which does not conflict with the other models but may act to reinforce the effect still further.

Their key finding, remember, was that epochs of *increased* magnetic field correspond to *colder* periods on Earth. Nobody knows why – perhaps the increased magnetic field funnels cosmic rays more efficiently on to the sensitive polar regions and that has something to do with it. But whatever the reason, they now wanted to find a way to modulate the Earth's magnetic field as its orbit around the Sun changes, in particular so that when the orbit is more elliptical the magnetic field is weaker and the Earth warms, tying in with the record of Milankovitch rhythms in the deep-sea cores. The magnetic field is produced by fluid motions in the Earth's electrically conducting core, so virtually all geophysicists believe, and it is strongest when the core fluid flows steadily, weaker when the core flow is disrupted.

Why should the flow be disrupted when the Earth's orbit around the Sun is more elliptical? Because the liquid core is more dense than the material outside it (the mantle), it is held together tightly by gravity and forms a more nearly perfect sphere than does the whole Earth. The Sun and Moon, tugging on both core and mantle, set up tidal imbalances between the two which might disrupt the core flow. In the case of the Moon, the changing patterns are regular and rhythmic, so that a stable flow soon gets established. The same is true of the Sun, provided the Earth's orbit is nearly spherical. But when the Earth's orbit is elliptical, the tidal effect of the Sun on the core and mantle is significantly different at one end of the orbit (December, say) from what it is at the other (July).

Could this changing influence be enough to disrupt the core flow, reduce the magnetic field and boost the warm part of the Milan-kovitch cycle? Somehow, I don't believe it. Everything hangs to-gether, but the resulting structure is rather contrived, and also rather redundant now that the sea-ice feedback details have been worked out. There is no reason why, with that feedback to help, the Milan-kovitch 100,000-year cycle shouldn't dominate the climatic patterns of the past 2 million years whatever the magnetic effect happens to be doing; indeed, it would be truly astonishing, in view of the complexity of the forces operating on the atmosphere, if the magnetic influence alone – or any other single influence – could explain all the changes we observe. The lesson here, though, is that while the wilder flights of fancy of the enthusiasts for a particular theory should always be taken with a pinch of salt, shooting them down doesn't invalidate all the ideas that come from the same source. The latest – and perhaps one of the best – ideas out of the Lamont stable came only after the initial excitement and controversy stirred by the earlier ideas had died down and Wollin, working on his own once again, had driven himself to the brink of collapse.

The short-term view

By the late 1970s other researchers had confirmed the reality of the magnetism/climate link on a time-scale of hundreds of thousands of years, and even without an entirely satisfactory mechanism to ex-plain the link the geophysicists seemed to have accepted that the case was proven. But Wollin wanted to establish that the relationship of weak field to warmer weather holds even on a time-scale relevant to forecasting in a human lifetime and could thus be a useful forecast-ing tool. His colleagues had returned to their own main lines of research, so it was in what he describes as a 'frenzied' solo effort to complete this work that Wollin plotted hundreds of curves to com-pare climatic data with magnetic declination, inclination and hori-zontal and vertical magnetic field intensities from different sites around the world. His depression at repeated failure became so grave that he became a virtual hermit, living in the guest-room of his own house and no longer communicating even with his own family. Then he read a review of the book *Essays of E. B. White*, in which the reviewer described how White cured a similar depression by 'work-ing with his hands and sipping sherry'.

Prepared to try anything once, Wollin made his way back to the

Lamont Observatory, where he scrounged junk from the workshops to make what he calls 'constructions'. (Hardly artistic enough to be described as sculptures, these piles of junk can be found scattered through the large wooded garden of his home to this day.) When he told his colleagues what he had in mind, Ryan went off to fetch the sherry. 'White is right,' Wollin told me. 'Working with your hands and sipping sherry does the job. In three weeks I had recovered from my depression, and there were four empty one-gallon sherry bottles in Eric's lab to take to the recycling plant.'

Eventually his renewed labours on the climatic problem bore fruit. The key to short-term change turned out to be not the intensity of the magnetic field, or any of its components, but the rate at which it was changing, either decreasing or increasing. This is clearly a different effect from the long-period links between magnetism and climate found by researchers around the world. Wollin had been barking up a wrong tree with his attempts to prove that the relationship of weak field to warm weather applies even from year to year and decade to decade. But what he did find is that the great sweep

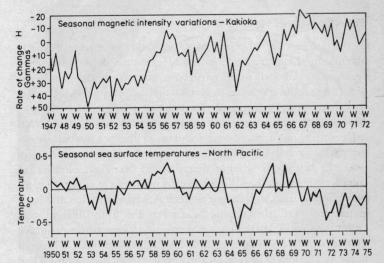

Figure 7.2. Over the past few decades, North Pacific sea surface temperatures correlate strikingly well with the rate at which the Earth's magnetic field has been changing – provided the temperature curve is shifted by three years to allow for the time it takes the atmosphere to respond to the changing magnetic field. Temperature changes are measured relative to an arbitrary baseline.

of ocean currents around the Pacific Ocean, the circulation of the biggest flow of water on Earth, is affected by the changing magnetic field.

Once he found the relationship, it was easy to see how it worked. The ocean water, laden with salt, is a moderately effective conductor of electricity, and its circulation sets up weak electric currents. These interact with the Earth's magnetic field through the dynamo effect.

The feedback between ocean currents (of both kinds) and the Earth's magnetic field is usually in equilibrium, but when the magnetic field changes rapidly it affects the ocean dynamo strongly enough, through the electromagnetic link, to alter the rate at which the great Pacific gyre sweeps around its ocean basin. By analogy with the dynamo effect, discovered by Michael Faraday back in the nineteenth century, a rapid increase in the magnetic field should

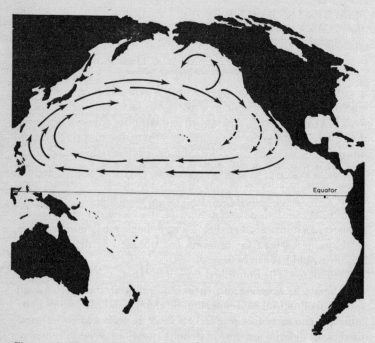

Figure 7.3. The links between Pacific temperatures and changing magnetism are explained by the dynamo effect of the great sweep of ocean currents around the Pacific basin, the North Pacific gyre.

have the opposite effect, allowing the circulation of the oceans to proceed vigorously and distributing heat more evenly over the globe. This is just the pattern of evens Wollin found, with the climate (temperature) changes at temperate and high latitudes following about three years after the sudden ups and downs of the magnetic field.

Partly because Wollin suffered a serious physical illness while completing this work, the results had not yet been published at the time this book was being prepared, but they should cause as much of a flurry as his papers of ten years ago when they do see the light of day. It seems there is a realistic basis here for predicting temperature fluctuations of half a degree or so above or below the long-term trend, in key areas of the globe for agriculture, three or four years ahead.

And that, tantalizingly, is where we have to leave the story of Goesta Wollin's investigation of the links between magnetism and climate. For anyone who thinks that, with the next Ice Age already due, warmer weather is probably a good thing, we can at last find one crumb of comfort from measurements of the Earth's magnetic field made by satellites during the 1970s. These show that the field is slowly decreasing in strength, at a rate of about 1 per cent per decade. If the trend continues, within a generation or so Wollin's scientific heirs may be able to make direct tests of the relationship: weak field/warm Earth. Apart from that crumb of comfort, however, the best evidence we have – volcanic activity, the historical record, sunspots and cosmic rays, the Milankovitch rhythms and all – indicate that the Earth is cooling, on a variety of time-scales but most importantly on a time-scale significant to mankind; the trend is back in the direction of the Little Ice Age after a bout of temporary warmth. Against that background of the changing climate, the prospect of a human-induced global greenhouse effect may not seem as alarming, at least for the next few decades, as the doomsday brigade would have us believe.

We shall see. The unfinished saga of the studies of geomagnetism and climate is, however, an entirely appropriate bridge between discussion of natural causes of climatic change and discussion of the anthropogenic greenhouse effect. Science is very seldom cut and dried, with definite answers to definite questions, especially where complex natural systems like the weather machine are concerned. When the interfering influence of people is added, understanding and predicting the way those natural systems will change becomes an almost impossible task which there is a pressing need to accomplish. If most of the scientists had their way, they'd prefer not even to discuss the greenhouse effect in public, but to work quietly on the problem for another twenty years or so before reaching any firm conclusions. Only the urgent need for planning and policy decisions on a global scale *if* the threat is real has brought them out into the arena of public debate.

In Part Two of this book I shall cover the best current understanding of the greenhouse effect and the problems it is likely to pose; but always bear in mind that as far as the research goes these are still early days, and the overall picture is still less clear-cut than the picture of how magnetism and weather interact (which is itself far from being a complete and rounded scientific theory yet) and much less clear-cut than the picture we now have of the most important climatic rhythms of the present epoch, the Milankovitch cycles. The greenhouse effect models are less than perfect. But they are the best we've got, and we need some basis for planning in a hungry world.

Part Two

The Global Greenhouse

8
Life and
the Terrestrial Greenhouse

The greenhouse effect has only recently made headlines, and the stories under those headlines often make it seem that the greenhouse effect is something new in the history of planet Earth, a unique side effect of human activities. But the greenhouse effect has actually been of key importance during the long history of our planet, a crucial contributor to the conditions which make life on Earth possible. Our sister planet Venus, only a little closer to the Sun than the Earth, is a superhot desert today because of a runaway natural greenhouse effect produced by its thick carbon dioxide atmosphere; the Earth is a water planet with a thin atmosphere, an ideal home for life. The differences depend entirely on the balance of the greenhouse effects operating on the two planets, and those differences can be traced back to the early days of the formation of the Solar System, more than 4 thousand million years ago. The manmade (anthropogenic) greenhouse effect is just a minor, passing phase on that sort of time-scale, no matter how important it may be for us.

But what, exactly, is the greenhouse effect? The Earth, like other planets, is warm primarily because it receives heat from the Sun. A little heat does escape from the interior of our planet, but this is tiny compared with the heat energy pouring down on us from the Sun. To reach the surface of the Earth, where we live, that solar energy has to pass through the atmosphere, and it does so with ease. Most of the Sun's energy is emitted in a band of wavelengths from 0·4 to 0·7 microns (one micron is a thousandth of a millimetre), called the visible band. Our eyes have evolved, over hundreds of millions of years, to make use of this radiant solar energy, so it is no surprise to find that most solar energy is radiated in what we call the visible band. The shortest of these wavelengths are violet in colour, the longest red, with the whole of the visible spectrum – the rainbow –

in between; altogether, about 46 per cent of the Sun's energy is radiated as visible light.

About 7 per cent of solar energy is radiated at shorter wavelengths, below 0·4 microns, called the ultra-violet. And the rest, long wavelength energy above 0·7 microns, is called infra-red, a term synonymous with radiant heat energy. When you hold your hands near a warm radiator and feel the heat, even though the radiator is not glowing visibly, it is infra-red radiation that you feel. It is just a coincidence that the Earth's atmosphere is transparent to the bulk of the Sun's radiant energy, in the visible band; on other planets, cloud layers high in the atmosphere block out incoming solar energy, which cannot penetrate to the depths. But life needs energy, so, once again, it is no surprise to find that we live on a planet whose atmosphere is transparent to most of the energy put out by the star it circles around. Nearly all the ultra-violet radiation from the Sun is, however, absorbed high in the atmosphere by molecules such as nitrogen oxides, oxygen, and ozone, the tri-atomic form of oxygen. This is just as well for life on the surface of the Earth, since ultra-violet radiation can damage living cells, breaking down the life molecule, DNA; the small amount of solar ultra-violet that does reach the ground causes sunburn and some forms of skin cancer.

Some of the incoming infra-red radiation is also absorbed in the atmosphere, chiefly by water vapour and carbon dioxide. The rest reaches ground level, along with almost all the visible radiation from the Sun, and warms the surface of the ocean, the land and the vegetation covering the land. The result is that the warm surface of the planet radiates energy back out into space. On average, the heat radiated out into space must exactly balance the heat arriving from the Sun; if, by magic, the Sun could be made a little hotter, then more energy would get through to the surface of the Earth, the planet would warm up slightly, and more heat would radiate out into space, with a new balance being struck. But because the Earth is so much cooler than the Sun, it radiates its energy at much longer wavelengths, almost entirely in the infra-red part of the spectrum from about 4 microns to beyond 30 microns. The energy around 8 to 12 microns escapes easily into space; but most of the rest is absorbed in the atmosphere by water vapour, carbon dioxide and traces of other gases. The result is that the atmosphere near the surface of the Earth warms up: this is the atmospheric greenhouse effect. Eventually the energy which warms the atmosphere does escape into space – the warm atmosphere itself radiates infra-red energy, some going up and out into space, some going back down to warm the ground and be re-radiated yet again. The two crucial effects of the atmosphere on radiant energy are that it absorbs incoming solar ultra-violet which would otherwise harm life on the surface of the Earth, and it holds in some of the outgoing infra-red, keeping the surface warm. But the atmosphere was not always like this.

The evolving atmosphere

The Earth and its neighbouring planets Venus (slightly closer to the Sun) and Mars (the next planet out from the Sun) formed as the Sun itself and the rest of the Solar System formed about $4\frac{1}{2}$ thousand million years ago. The planets grew out of a disc of gas and dust circling the young Sun, something like the rings around Saturn today, but on a much bigger scale. Particles of dust in the ring system collided and stuck together, building up into bigger and bigger lumps orbiting around the Sun. Some of these aggregates grew so big that their gravitational pull began to dominate the region of space around each of their orbits, pulling in the other rocky lumps and building up into true planets. As the growing planets swept the space around them clean, they were bombarded with meteorites; the battered faces of the Moon and Mercury today show what a pounding all the planets took in those far-off days; on Earth (and Venus and, to a lesser extent, Mars) the craters have been softened and smoothed away by erosion.

Each of the young planets was hot, a ball of molten rock warmed by the release of kinetic energy (energy of motion) as the meteorites struck, by the release of gravitational energy as the rocky ball squeezed down into a sphere under the tug of its own gravity, and by the release of nuclear energy as radioactive heavy elements fissioned into lighter elements in its interior. But very early on in the history of the Solar System any traces of gas around the young planets were swept away by the violently erratic outbursts of the young Sun before it settled down. The present-day atmospheres of the Earth, Mars and Venus must have been produced as gases were released from their interiors by volcanic activity and by the vapourization produced by the impact of meteorites. There is no reason to think that the mixture of gases produced by outgassing from the Earth's interior should be significantly different now from the mixture outgassed earlier in the history of our planet, so geologists and astronomers try to get an understanding of the early atmosphere of the Earth (and the similar planets nearby – the terrestrial planets) by looking at the mixture of gases released by volcanoes today.

Surprisingly, in view of their association with fire, the main 'gas' that volcanoes release is water vapour. Different styles of eruption produce different mixes of gas. For obvious reasons, most measurements have been made for the quieter eruptions of the Hawaiian type – it is rather difficult to make direct measurements of the gas released by an explosive volcano like Mount St Helens. But the Hawaiian figures probably do not differ very much from the global average. They show that, in terms of weight, 64 per cent of the volatile material produced by volcanoes is water vapour, 24 per cent is carbon dioxide, sulphur makes up almost 10 per cent, and nitrogen just over 1½ per cent. The puzzle is how that mixture, leaking from volcanoes over thousands of millions of years, has produced an atmosphere which is today 78 per cent nitrogen, 21 per cent oxygen,

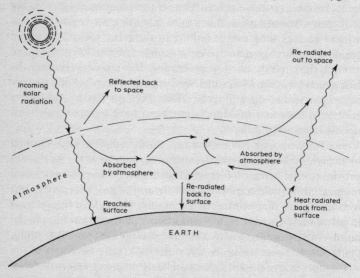

Figure 8.1. The greenhouse effect. About 40 per cent of incoming solar radiation is reflected back to space, 15 per cent is absorbed by the atmosphere, and about 45 per cent reaches the Earth's surface. This is eventually re-radiated as heat (infra-red) radiation. Outgoing radiation absorbed in the atmosphere and re-radiated back to the surface keeps the Earth warmer than it would be without an atmosphere.

with only a trace of carbon dioxide and about 1 per cent other gases. Equally, why does Venus have an atmosphere containing 97 per cent carbon dioxide, and Mars one containing 95 per cent carbon dioxide – figures much more plausible in terms of the volcanic mixture of gases? The answers depend upon a combination of the greenhouse effect and the presence of life on Earth.

When the hot rocks of the Earth's surface cooled below the boiling point of water, 100° C, the enormous quantities of water vapour present in the early atmosphere, and being outgassed by volcanoes, began to fall as rain. The geological evidence shows that by 3.8 billion years ago there was liquid water on the surface of the Earth – our planet became the water planet very early in its history. The oceans of water did two things. They provided a permanent source of moisture for the atmosphere, ensuring a regular hydrological cycle and keeping a large part of our planet covered by white clouds. And they dissolved great quantities of carbon dioxide, providing an escape route for carbon dioxide from the atmosphere, eventually to be laid down in sedimentary rocks such as limestone, rich in calcium carbonate. The result was that a thick, hot atmosphere rich in water vapour was turned into a global ocean and a thinner, cooler atmosphere still dominated by carbon dioxide.

On Venus, things were rather different. To start with, because the ring around the Sun from which Venus formed was that much closer to our parent star (41 million km closer than the Earth) it probably contained less water from the outset, because this would have been driven outwards by the heat of the Sun. Secondly, as the young Venus began to cool, its surface temperature would have levelled off at a higher point than the temperature on Earth because of its proximity to the Sun. Setting the Earth–Sun distance as one astronomical unit, the distance of Venus from the Sun averages out at 0.72 units, and that of Mars at 1.52 units. The rule of thumb used in calculating the temperature of an *airless* planet is that this goes as one over the square root of its distance from the Sun. So, other things being equal, the temperature of Venus would have been $1/\sqrt{(0.72)}$ = 1.2 times that of the Earth, and similarly the temperature on Mars ought to have been 0.8 times that of the Earth – in round terms, Venus 20 per cent warmer than our planet and Mars 20 per cent cooler. The actual temperature on Earth today averages out at about 15° C; on Venus the temperature is a searing 500° C, and on Mars it ranges from about − 30° C down to a frigid − 140° C at the poles in winter. The *percentage* differences only apply to temperatures measured on the absolute or Kelvin scale, and these come out as 288 K for the Earth, beyond 700 K for Venus, and 230 K for Mars.* Mars is, more or less, the 'right' amount cooler than the Earth, but Venus is superhot. The deficiency of water in its early atmosphere and the slight extra burden of solar heat were enough to ensure that from the early days of the Solar System Venus retained a thick carbon dioxide blanket around itself, a blanket which has trapped infra-red radiation and allowed surface temperatures to soar by a runaway greenhouse effect.

Without some dramatic change in the Sun's output this could never happen on Earth. Here, there is an inbuilt 'thermostat' which has maintained temperatures close to an average 15° C for 3 billion years. If the Sun were a little hotter more heat would reach the oceans and more water would evaporate, making more shiny white clouds. The white clouds would reflect some of the extra solar heat into space, helping to keep the Earth cool. On the other hand, if the Sun cooled a little then there would be less evaporation, fewer clouds in the sky, and a greater proportion of the reduced solar energy would reach the ground, keeping it warm. The oceans provide a buffer against extremes of both heat and cold – but this buffer operates on a geological time-scale, not a human time-scale, so we cannot assume it will protect us from the anthropogenic greenhouse effect in the next century or so. We can, though, be sure that no matter how much fossil fuel we burn, no matter how complete our destruction of tropical forests may be, the Earth will not, through a runaway greenhouse effect, come to resemble the desert planet Venus.

The stability of the Earth's temperature for 3 billion years or more, revealed unambiguously by geological evidence for the presence of liquid water throughout the long history of our planet, is even more remarkable because the Sun's output has indeed changed in that time. Astronomers calculate that the output of energy by the Sun has increased by some 30 per cent over the past 3 billion years, yet the Earth was not a frigid, ice-covered ball 3 billion years ago. The only explanation for this is that the greenhouse effect was at work then, keeping the surface of the Earth warmer than it would otherwise

---

* The conversion from Kelvin to degrees Centigrade simply involves subtracting 273.

have been. Planetary scientists are not sure exactly how the effect worked that long ago, but very probably, in addition to carbon dioxide, a secondary influence was the presence of ammonia in the early atmosphere. Ammonia is a very good greenhouse gas, happily absorbing infra-red radiation from the ground, and it is one of the minor constituents of the volcanic gas mixture – one of the main sources of that $1\frac{1}{2}$ per cent of nitrogen, each nitrogen atom being locked together with three hydrogen atoms as ammonia, NH_3.

Ammonia almost certainly played a part in another dramatic event in the history of the Earth, the appearance of life. Many experiments have now shown that a mixture of water, carbon dioxide, a little ammonia and traces of other compounds produced by volcanic activity can be energized by ultra-violet radiation (simulating the early Sun) or by electric sparks (simulating lightning in the early atmosphere of the Earth) to produce a sticky, dark brown goo which is rich in the kind of organic molecules that are thought to be the precursors of life. Although nobody knows exactly how life got a grip on Earth, there is nothing surprising in the fact that it did so on a wet planet with a carbon dioxide atmosphere laced with a little ammonia. From then on, the story of the evolving atmosphere and the evolution of life are part and parcel of the same tale.

Life and oxygen

Venus, a little closer to the Sun than we are, became the overheated victim of a runaway greenhouse effect. On Mars, a little further from the Sun and too small for its gravity to hold on to a thick atmosphere, the water froze and the planet became a cold desert. But on Earth the early carbon dioxide atmosphere helped to keep the planet warm enough for oceans of liquid water to exist, and in those oceans life began to evolve in conditions that were, like Baby Bear's porridge in the Goldilocks story, 'just right'. Life in the oceans then in its turn modified the atmosphere of the Earth until it began to resemble the atmosphere we know today.

The presence of all that nitrogen today is not very surprising at all, since it does not readily form stable compounds with other elements and tends to stick around once it is released into the atmosphere. Compounds such as ammonia are easily stripped of their hydrogen by chemical processes – especially biochemical processes – leaving nitrogen behind, and over 3 billion years or more that $1\frac{1}{2}$ per cent produced by outgassing had ample time to build up. With the water condensed as oceans and most of the carbon dioxide locked up in sedimentary rocks, the only major change in the atmosphere of our planet in the past 3 billion years has been the conversion of the remaining carbon dioxide into oxygen. But that change is all-important.

Plants are chemical factories that make glucose, a source of energy for living things, and oxygen, a waste product, out of carbon dioxide and water. This process is known as photosynthesis, because it can take place only in the presence of light (photons). Unlike their modern descendants, the first photosynthesizers, dominating the oceans of the world from 3 billion years ago until less than 2 billion years ago, did not release the oxygen they produced into the atmosphere. To them, oxygen was a deadly poison which could only safely be released locked up in compounds with iron. The result was the production of layers of iron oxides, known as Banded Iron Formations, or BIFs, found around the world in geological strata between just over 3 and $1\frac{1}{2}$ billion years old. But between 1·8 and 1·5 billion years ago the deposits of BIF ceased, as new forms of photosynthesizing life arose that were able to live with free oxygen. Instead of having to go through the energetically wasteful process of locking up oxygen with iron, the new life forms released it into the atmosphere, gaining an incidental bonus in the struggle for survival as the free oxygen killed off many of their competitors. As free oxygen built up, much more highly oxidized deposits of iron were laid down as red beds – the Earth literally rusted – and increasing amounts of carbon dioxide were locked up as carbonates, compounds which include extra oxygen as well as carbon dioxide.

Increasing quantities of oxygen in the atmosphere from about 1 billion years ago were involved in reactions with sunlight (photochemical reactions) that built up the layer of ozone (tri-atomic oxygen) in the stratosphere. The ozone layer then blocked incoming solar ultra-violet radiation, and only from this time onwards was there any possibility of life moving out of the sea and on to the land. At the same time, free oxygen in the air opened the way for the evolution of animal life, which breathes in oxygen and uses it to metabolize food and provide energy. Without the photosynthesizers, there would be no ozone layer to shield us from solar ultra-violet, and no oxygen to breathe.

The explosion of fossil remains of complex life forms, the beginning of the Cambrian period of geological time, coincides (if that is the right word!) with the establishment of atmospheric conditions very similar to those of the present day, about 600 million years ago. Since then, life has diversified and moved on to the land under the shield of the ozone layer, breathing an atmosphere which has remained 78 per cent nitrogen and 21 per cent oxygen, with just a trace of everything else, including carbon dioxide. Today, that trace of carbon dioxide makes up about 0·03 per cent of the dry weight of the atmosphere. Some geological evidence suggests that in the past 600 million years that concentration may have varied, through natural causes, between about 0·02 per cent and 0·04 per cent; this is a modest range of fluctuation, but it opens the way for interesting speculations about whether, for example, a slightly more carbon-dioxide-rich atmosphere might have contributed both to the warmth of the Carboniferous period around 300 million years ago (through the greenhouse effect) and to the growth of the vegetation on which the reptiles of the time fed (through the availability of more carbon dioxide for photosynthesis by plants). These are not entirely idle speculations, since it is the fossilized remains of those Carboniferous forests, in the form of coal, that we are burning today, thereby causing a build-up of carbon dioxide in the atmosphere.

The palaeoecologists cannot yet give detailed answers to these speculative questions, but they do make the important point that even by burning all available fossil fuel we would be doing no more than putting back into the atmosphere carbon dioxide that used to be there long ago but has been sequestered by plants through photosynthesis. Human efforts cannot, from that perspective, create any conditions that have not occurred naturally during the past 600 million years; we are *not* heading for a runaway greenhouse or the end of the world as a fit home for life, merely at worst a minor, short-lived variation which has little significance in the $4\frac{1}{2}$-billion-year history of our planet. Any problems that may arise for us through the anthropogenic greenhouse effect are simply human problems – the temporary problems of a species which has itself been around for only a tiny fraction of our planet's history. That may be comforting for anyone broadminded enough to have the good of life on Earth more at heart than the good of the single species of which we are members. But from the viewpoint of most members of that species, humanity, it leaves us with plenty to worry about here and now. Oxygen and nitrogen do not absorb infra-red radiation significantly, and so do not contribute to the global greenhouse. Carbon dioxide does absorb infra-red, so any build-up of carbon dioxide is a cause for concern.

The build-up of anthropogenic carbon dioxide

The present concentration of carbon dioxide in the atmosphere, about 0·03 per cent, is commonly expressed as 300 parts per million (ppm). The current debate about the greenhouse effect centres on recent changes in this figure, changes which have almost certainly been brought about by human activities. When fossil fuel is burnt, 'old' carbon is put back into the atmosphere as carbon dioxide. This involves removing some oxygen from the atmosphere, but the amount of oxygen present, 21 per cent of the atmosphere, is so enormous compared with the amount of carbon being burnt and the amounts of carbon dioxide being made, a few parts per million, that this poses no conceivable threat to the store of oxygen we need to breathe. To put this in perspective, the atmosphere contains 1,200 trillion (million million) tonnes of oxygen, and just 2·6 billion (thousand million) tonnes of carbon dioxide; if all plant life – including the single-celled life in the oceans – died tomorrow and photosynthesis ceased, we would all starve very quickly and the Earth would be lifeless. But it would take millions of years – an eyeblink on the geological timescale – for all that oxygen to be combined with other elements, notably iron, carbon and sulphur, and for the Earth's atmosphere to adjust to the absence of life. Suffocation is very much *not* a part of the greenhouse-effect problem, although, like the equally mythical threat of a 'runaway greenhouse', it has been aired in the wild-eyed theorizing of the more extreme doomsayers.

There is no doubt that the carbon dioxide concentration in the atmosphere is currently increasing, and very little doubt that for the past 20 years at least this increase has been predominantly due to the burning of fossil fuels. The build-up of carbon dioxide probably dates back to the industrial revolution in Europe at the end of the eighteenth century, but there are no direct measurements of the carbon dioxide concentration from more than about a hundred years ago, and even the measurements that are available from the 1880s are uncertain and sometimes contradictory. One way to try to get a handle on the build-up of carbon dioxide in the nineteenth century is to estimate how much has been produced by human activities.

Against the background of these activities, it is worth remembering that about 100,000 times as much carbon dioxide as there is in the atmosphere today has been released from volcanoes over geological time and is now locked up in rocks such as limestone and dolomites. The amount of carbon dioxide stored in the sedimentary rocks of the Earth is almost exactly the same as the amount present in the atmosphere of Venus, as measured by recent space probes – happy confirmation of the role of volcanic outgassing on each of these two similarly sized planets. One human activity, the manufacture of cement, does release some of this primordial carbon dioxide back into the atmosphere, as lime is extracted from limestone. But cement production today is only around 500 million tonnes per year (0·5 Gigatonnes, or 0·5 Gt) and this releases only about 0·5 Gt of carbon dioxide each year. Impressive though that figure is by everyday standards, it is, as we shall see, only about 3 per cent of the amount of carbon dioxide currently released by burning fossil fuels, and it can safely be ignored in the present debate about the greenhouse effect. For all practical purposes, the carbonate rocks provide a permanent deposit of carbon dioxide in the Earth's crust.

That leaves four reservoirs between which carbon dioxide is continually being exchanged in a dynamic equilibrium. Fossil fuels are deposited over the course of geological time, removing carbon from the carbon dioxide cycle. They are burnt by us on a much more rapid time-scale, putting carbon dioxide back into circulation. The atmosphere provides the reservoir for carbon dioxide which matters most in terms of the greenhouse effect, while some is dissolved in the oceans and some carbon is stored in the living plants and animals (chiefly plants) of our planet – the biomass.

When 1 Gt of carbon is burnt – as coal, say – it releases roughly 4 Gt of carbon dioxide, as each carbon atom combines with two oxygen atoms from the air. Between 1850 and 1950 about 60 Gt of carbon was burnt as fossil fuel, releasing about 240 Gt of carbon dioxide to the air. And by the late 1970s the *annual* input of carbon dioxide to the atmosphere from burning fossil fuels had reached 20 Gt. However, only about half of this carbon dioxide seems to stay in the atmosphere today.

The atmosphere itself contains about 700 Gt of carbon in the form of carbon dioxide (that is, some 2,800 Gt of carbon dioxide), and the living biomass contains about 800 Gt, slightly more than the atmosphere. Organic remains such as humus and peat probably contain at least 1,000 Gt of carbon, and some estimates range up to three times that amount – there is considerable uncertainty and disagreement even among the experts. The 50 per cent of carbon dioxide from burning fuel which does not seem to be staying in the atmosphere today presumably dissolves in the oceans, where carbon dioxide interacts with dissolved carbonates to form the bicarbonate ion. This dissolved reservoir contains about 40,000 Gt of carbon, but only about 600 Gt of this is in the surface layers of the ocean where it can easily interact with the atmosphere. The best guess at the amount of carbon stored in the ocean as organic remains, the oceanic equivalent of humus, is 3,000 Gt, and against all this the estimate of the total amount of carbon stored in fossil fuels is a relatively modest 12,000 Gt. Perhaps 60 per cent of this might be recoverable by mankind and burnt; yet again we see that human activities are not causing drastic changes to the environment, merely putting back into circulation a minor proportion of carbon that has been temporarily locked away.

In the long term – 1,000 years or more – the oceans are capable of absorbing 80 per cent or more of the carbon dioxide released by burning fossil fuel. But the fuel is being burnt more rapidly than the oceans can dissolve the end product, and in the decades and centuries ahead the build-up of carbon dioxide could pose a temporary greenhouse problem while the oceans are adjusting.

Accurate figures for the build-up of carbon dioxide in the atmosphere go back only to 1957, when a monitoring programme, run by the Scripps Institution of Oceanography, began at a site on the Mauna Loa mountain in Hawaii. This site in the Pacific Ocean is far from any sources of industrial pollution and ideal for monitoring large-scale changes in the atmosphere. The Mauna Loa measurements show a steady rise in carbon dioxide concentration for the past 25 years (Figure 8.2). The originally quoted figure for the carbon dioxide concentration in 1957 was 311 ppm; Bert Bolin of the University of Stockholm suggests that with a more accurate calibration of the equipment this baseline figure should be taken as 315 ppm. Either way, the build-up to above 335 ppm by 1980 is impressive. But problems come when we try to trace this build-up back into the nineteenth century.

Apart from the known amount of fossil fuel burnt, there is a way to estimate the amount of carbon dioxide in the atmosphere in the past century using the abundances of different isotopes of carbon found in the wood of tree-rings from different years. As well as the

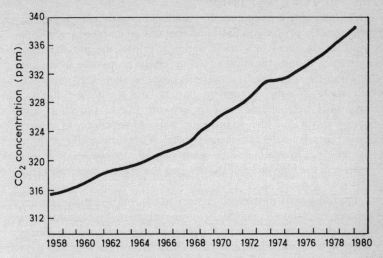

Figure 8.2. With the seasonal variations removed, the Mauna Loa measurements of carbon dioxide in the atmosphere strikingly demonstrate the build-up of this greenhouse-effect gas over the past three decades.

long-term trend, the Mauna Loa records show a seasonal fluctuation in carbon dioxide, ranging up and down by 5 ppm, which is explained as the effect of plants taking in carbon dioxide during Northern Hemisphere spring and summer and releasing it as they drop their leaves and decay in the autumn and winter. Growing plants take up the isotope carbon-12 in preference to carbon-13, so the proportion of carbon-13 is lower in the biomass than in the atmosphere; modern techniques are sensitive enough to measure the change in the ratio of the two isotopes in atmospheric carbon dioxide at Mauna Loa over the year, confirming that the seasonal variations are the result of the biomass 'breathing'.

The role of the forests

Similar techniques applied to tree-ring studies ought to reveal the proportion of carbon from fossil fuel and burnt wood present in the atmosphere as carbon dioxide each year during the nineteenth century. As well as carbon-12 and carbon-13, living things contain radioactive carbon, carbon-14, made by the interaction of cosmic rays with nitrogen in the atmosphere. But carbon-14 decays with a half-life of only a few thousand years, and coal deposits are many millions of years old. So fossil fuel contains no carbon-14 at all. As fossil fuel is burnt over the decades, it dilutes the proportion of carbon-14 in the atmosphere, and hence the proportion being laid down in tree-rings; as fresh wood is burnt, it alters the proportion of carbon-12 in the atmosphere, putting back the isotope that had previously been selected out by growing plants; this effect also happens with fossil fuel. In principle, it should be possible to unravel these effects from delicate measurements of the carbon isotope ratios in tree-rings from nineteenth-century wood, and Professor Minze Stuiver of the University of Washington has made a bold stab at the task.

From the early nineteenth century to the mid twentieth century the tree-rings show a gradual decline in carbon-13. Stuiver interprets this as due to the increase in the proportion of carbon-12 as forests were cut back and burnt, and calculates that between 1850 and 1950 120 Gt of carbon were released from the biomass. This is twice the amount of carbon burnt as fossil fuel over the same interval, and Stuiver's somewhat controversial conclusion is that forest clearance dominated the build-up of carbon dioxide in the atmosphere until the middle of this century. His figures imply a bigger growth rate for the build-up than is suggested by any other expert, and indicate that the baseline from which this growth took off in 1850 was only about 270 ppm. Late nineteenth-century measurements of the carbon dioxide concentration come out rather higher than Stuiver's estimates based on tree-rings, and most authorities favour those figures, arguing that Stuiver has made a mistake somewhere with his isotope technique and that the concentration of carbon dioxide in the atmosphere may already have been as high as 290 ppm in the middle to late nineteenth century. This is an area of great uncertainty, but it is highly important for the debate about the build-up of carbon dioxide in the atmosphere because, paradoxically, if Stuiver is correct then it seems the oceans are even more efficient at absorbing carbon dioxide than we thought, and we may therefore have *less* to worry about than might otherwise be the case.

Stuiver's suggestion that twice as much carbon dioxide was produced by destruction of forests between 1850 and 1950 as from burning fossil fuels ties in with at least two other studies. One, by the New Zealander A. T. Wilson, calculates that, far from being a steady build-up of carbon dioxide, most of it occurred in a great burst between 1860 and 1890. Wilson, who also uses tree-ring studies in his analysis, links this burst of carbon dioxide production with the explosion of pioneer agricultural activity and forest clearance at temperate latitudes which followed the opening up of virgin lands in North America, New Zealand, Australia, South Africa and Eastern Europe by the railways. If Wilson is correct, the biomass contributed in one 30-year interval as much carbon dioxide as all the fossil fuel burnt between 1850 and 1950. He even suggests that such a dramatic change, increasing the carbon dioxide content of the atmosphere by 10 per cent in the space of a few decades, may have been a major factor in the roughly 0·5° C rise in global mean temperatures which pulled us out of the Little Ice Age and into the more comfortable weather of the twentieth century. Very few climatologists or isotope experts support Wilson's conclusions; right or wrong, though, they serve as a timely reminder that greenhouse effects and global warmings may not be all bad. Most people would agree that the climate today is an improvement on Little Ice Age conditions, and whatever the validity of the 'pioneer agriculture explosion' hypothesis, the pleasant weather of the twentieth century may be maintained at least in part by the anthropogenic greenhouse effect.

The other study which ties in with Stuiver's work brings this kind of investigation bang up to date. If destruction of forests may have dominated the build-up of carbon dioxide in the nineteenth century, how much might they be contributing to the present day build-up, measured at about 2 ppm per year? In 1977 and 1978 George Woodwell of the Marine Biological Laboratory at Woods Hole, Massachusetts, and Bert Bolin in Stockholm gave a new slant to the whole carbon dioxide debate.

Measurements at sites such as Mauna Loa currently show a build-up of carbon dioxide roughly equivalent to the release of 3 Gt of carbon, or slightly less, each year. But in 1978 and 1979 the amount of fossil fuel burnt was equivalent to 5·1 and 5·4 Gt of carbon, respectively. Rather more than 2 Gt of carbon – in some years, more than 2½ Gt – is being taken up by the oceans or other natural processes, perhaps in the formation of limestone rocks. Oceanographers and ocean chemists say that the oceans cannot be absorbing all of this surplus, and until Woodwell and Bolin upset the applecart the standard argument was that a good proportion of the extra carbon dioxide was being taken up by the biomass – forests were stimulated into increased growth by the availability of more carbon dioxide for photosynthesis. Now it seems that the forests too may be producing a net input of carbon dioxide to the atmosphere, exacerbating the problem. Woodwell and Bolin independently used estimates of the amount of forest being destroyed each year to come up with a figure for the amount of carbon dioxide being put into the atmosphere from the biomass. Bolin's estimate was about 1 Gt a year, Woodwell's about 2 Gt a year – sufficiently close to suggest that they were both somewhere near the truth. Other more recent studies have come to broadly similar conclusions, and today there is a general agreement that the biomass could well be a small net contributor of carbon dioxide, but that its contribution is less significant than the 5 Gt or more of carbon released from fossil fuel each year.

The important conclusion from all this work is that 'sinks' for carbon dioxide are more efficient than current theories can explain. Somehow the oceans – or something else – are absorbing more carbon dioxide than we can account for. But there is no guarantee that they will continue to absorb this much as the amount of carbon dioxide in the atmosphere continues to build up. An optimist might argue that if the sinks are so efficient, an increase of carbon dioxide might stimulate them into even more efficient activity; a pessimist would say that if they have been absorbing so much carbon dioxide they must by now be nearly full up, and will be unable to cope with the products of increased burning of fossil fuels. Caught between the two extremes, most of the contributors to the carbon dioxide debate today can only guess at the proportion of the carbon dioxide produced by human activities that is likely to stay in the atmosphere in the years and decades ahead. We know from direct measurements that for 20 years roughly half of the carbon dioxide released by burning fossil fuels has stayed to build up in the atmosphere. The reasonable guess is that biological factors will stay the same for the next 20 years, and that all the sinks will continue to operate as before. So the rule of thumb used in climatic forecasting is that roughly half of the carbon dioxide released from burning fossil fuel in the next 20 to 50 years will stay to build up in the atmosphere.

This is no more than a guess based on past experience. It could very well prove wrong – either way – but scientists today have no basis for making any better guess. They would prefer not to guess at all, but if they wait until their understanding of the carbon cycle is perfect, it will be too late for society to plan for the changing conditions – they will already have changed. Planners need forecasts, and even crude forecasts will do if they are the best we have got. But there is no point in pretending that they are better than they really are.

Measurements of the carbon dioxide concentration in air bubbles trapped in the polar ice-sheets show that during the most recent Ice Age the carbon dioxide concentration of the atmosphere may have been only half what it is now; that is probably because cold ocean water dissolves carbon dioxide more effectively. Whatever our guesses about the future trend of the carbon dioxide build-up, it is clear on this evidence that we are heading for a situation unique in the past 100,000 years of Earth history. In fact, in view of the pattern of recent glaciations, and the way the world's weather has gone for the past few million years, it seems likely that we are heading for a situation unique in the past 2 *million* years of Earth history. Forecasting the consequences is bound to be a matter of guesses, some more inspired than others. The task confronting the forecasters is first to guess how much fossil fuel will be burnt over the next 50 years or so, and how quickly (a forecasting task that has defied the collective wisdom of more than one think tank!), and then to predict how the resulting carbon dioxide build-up will increase global temperatures through the greenhouse effect and thereby alter the pattern of climate around the globe. Only a fool would take the resulting 'guesstimates' as gospel truth about what lies ahead; but only a bigger fool would ignore the problem entirely in the hope that it may go away of its own accord.

9
Carbon Dioxide and Climate: Business as Usual?

The simplest way to get a first impression of how serious the carbon dioxide problem may be is to assume that economic activity and energy use in the world continue to develop without any thought being given to the question of carbon dioxide. It ought then to be possible to estimate the effect of the resulting build-up of carbon dioxide on climate. This is the 'business as usual' scenario – but, unfortunately, nobody today can be quite sure what this actually means. In the early 1970s, and even after the first oil crisis, many forecasters talked of continuing growth in energy demand at a rate of 4 per cent per year, or even more, into the twenty-first century; at the other extreme, by the late 1970s economic pessimists were arguing that an era of zero growth was upon us whether we liked it or not. The 'business as usual' scenario makes no judgements about whether growth is good or bad, it simply tries to assess what will happen if natural trends are left to take their course – then, if the likely outcome appears undesirable, it may be that action will be taken to ensure that such an outcome never happens. Such scenarios are not 'forecasts', in the sense that they are predictions of what *will* happen; they are images of *possible* future worlds from which, perhaps, we may be able to select the most desirable possibilities.

As well as the economic uncertainties that make forecasting future energy use, in the sense of a definite prediction, virtually impossible today, there are many areas of uncertainty in the greenhouse-effect calculations themselves. Professor Kondratyev, whom we met in Chapter 5, is one of the climatologists who stress that there are other polluting gases, besides carbon dioxide, which contribute to the greenhouse effect. These include the fluorocarbons from spray cans which have gained notoriety through the possibility that they may damage the ozone layer, oxides of nitrogen, sulphur dioxide and water vapour. Of these, water vapour may be of crucial importance and is included in the best greenhouse-effect calculations. As for the others, it is impossible to forecast how they may build up in the atmosphere over, say, the next 50 years, so it is impossible to calculate even approximately their contribution to the greenhouse effect. Carbon dioxide is certainly the major component of the problem today, but even when the carbon dioxide greenhouse effect is fully understood, there will still be plenty more work for the climate modellers to do. Meanwhile, we have to make do with that approximate first cut at the problem.

Possible energy futures

There is so much fossil fuel available, mainly in the form of coal, that there is no prospect of the world running out of coal before the carbon dioxide build-up becomes important. The 'energy crisis' that we hear so much about is related to increasing *costs* of energy, rather than any real shortage of fuel, and to the likelihood that oil supplies may begin to run short at the end of the century. Oil is a very convenient, easily transported and (until recently) cheap provider of energy. Coal is dirtier, harder to extract from the ground, more

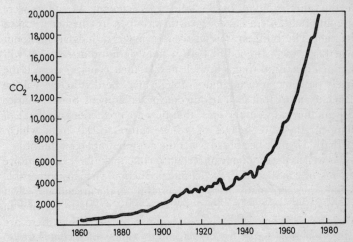

Figure 9.1. Calculations of the amount of carbon dioxide released to the atmosphere by burning fossil fuel show how the build-up is reaching runaway proportions. The figures are in million tonnes, and exclude carbon dioxide from forest clearance. (Source: *Earthscan*.)

difficult to transport, and (until recently) more expensive. But when and if the oil runs out, we can run on coal, if need be (and if the greenhouse effect allows), for hundreds of years while alternative sources of energy are developed. The key to the build-up of carbon dioxide in the atmosphere is not the amount of fossil fuel there is in the ground, but how quickly we dig it up and burn it.

Geologists estimate that the total recoverable reserves of fossil fuel in the world, 90 per cent of it coal, are equivalent to energy amounting to 300,000 quads. One quad is one thousand million million British Thermal Units, one quad per year the equivalent of the output of twenty 1,000 megawatt power stations. One quad per year is the present-day energy consumption of each 10 million inhabitants of the USA.

At present the whole world uses 250 quads per year, and at this

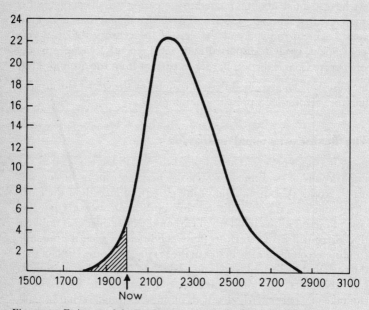

Figure 9.2. Estimates of the Earth's total fossil fuel reserves indicate that so far we have burnt only a small fraction (shaded area) of the available carbon reserves. With unrestricted growth, annual consumption might peak in about 2200 at five times present levels, before falling off as supplies run out. Fortunately, this is not a realistic scenario. (Source: Earthscan. Vertical scale in gigatonnes – thousand million tonnes.)

rate the known recoverable reserves of fossil fuel would last for another 1,200 years. Any growth in use of fossil fuel shortens this time-scale and also leads to a more rapid build-up of carbon dioxide in the atmosphere, giving the oceans and other systems that absorb carbon dioxide less time to adjust and to remove the excess from the atmosphere. In the early 1970s the global consumption of fossil fuel was increasing at a rate of slightly more than 4 per cent a year, equivalent to doubling the amount of fossil fuel burnt every 16 years or so. The available reserves could easily supply such rapidly growing demand for another hundred years or so, but would then be rapidly depleted. Since 1974, with increased energy costs and a worldwide economic recession, the increase in consumption of fossil fuel has been a slightly more modest $2\frac{1}{2}$ per cent. In very round terms, the 4-per-cent growth rate would have meant a doubling of the pre-industrial carbon dioxide concentration, to about 600 ppm, by the year 2030; reducing the growth rate to 2 per cent only pushes the date of the carbon dioxide doubling out to the year 2070 (both of these scenarios assume that half of the carbon dioxide produced continues to be absorbed by the natural sinks). Just as important as the global totals, however, are the likely ways in which patterns of fuel use will change in different regions of the globe during the lifetimes of most of its present inhabitants.

The 'business as usual' scenarios

Several groups of researchers around the world have looked at the implications of continued growth of energy use in a 'business as usual' world. A continuation of the pattern of growth of the early 1970s almost certainly indicates an upper limit, a 'worst possible case', in terms of the build-up of carbon dioxide, since it is unlikely that there will be a return to those conditions – even if economic conditions change for the better in the very near future, the fact that the carbon dioxide build-up is now widely recognized as a problem will surely affect planning in at least some parts of the world. So we can take one particularly detailed 'business as usual' scenario, published in 1978 by Ralph M. Rotty, of the Institute for Energy Analysis, Oak Ridge Associated Universities, as typifying the kind of calculation which began to make people sit up and take notice of the carbon dioxide problem.

Rotty divided the world into six sectors and projected the possible growth of fossil fuel use in each sector – not just the world as a whole – to the year 2025. He started his calculations from a baseline of 1974 energy use and assumed:

1. Zero energy growth in the USA, with consumption steady at 125 quads, 15 per cent of this non-fossil.

2. Growth at 2 per cent a year in Western Europe, with 15 per cent of this non-fossil.

3. Growth at 4 per cent a year in centrally planned Europe, including the USSR, again with 15 per cent non-fossil.

4. Similar growth in Western Europe, Japan and Australia.

5. Growth in China and centrally planned Asia at $4\frac{1}{2}$ per cent a year, almost all of this fossil fuel.

6. An annual *population* growth of 1·5 per cent a year in the rest of the 'Third World', accompanied by sufficient increase in energy use to ensure that by 2025 consumption per head had reached the global average of 1970. This involves roughly 5 per cent annual energy growth for this region.

These figures represent only one scenario, and anyone reading them will no doubt find some of them more plausible than others. Other scenario painters use slightly different figures and make slightly different assumptions, but all of them come up with much the same broad picture. In the case of Rotty's chosen numbers, the scenario projects a global energy demand in 2025 of 1,090 quads of fossil fuel, compared with 250 quads in 1980. This is a typical figure, close to the estimates made by other people using the same sort of 'business as usual' approach. And, again producing a scenario similar to those developed by other people, Rotty's calculations show the great shift of emphasis from the present developed world (the USA, Europe and the USSR) to the Third World over the next few decades. By the early twenty-first century, according to any sort of 'business as usual' picture, by far and away the biggest users of fossil fuel, and producers of carbon dioxide, will be the countries of the present-day poor, less developed parts of the world (see Figure 9.3).

This realization is crucially important in any discussion of the carbon dioxide greenhouse effect and the climatic problems it may pose. The people who are aware of the problem and might be motivated to do something about it are, by and large, the people of the developed world – the rich North. But the main producers of carbon dioxide in the near future will be the countries of the poor

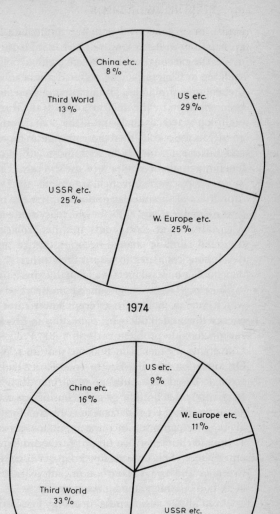

Figure 9.3. Global CO$_2$ production. This 'business as usual' energy scenario assumes zero energy growth in the USA between now and 2025, 2 per cent growth in Europe, 4 per cent in the USSR and Eastern Europe, 4·5 per cent in China and 1·5 per cent in the Third World. On this basis, making likely assumptions about the fossil fuel use of each region, CO$_2$ emissions by 2025 will be six times those of 1974. But the Third World and China, which are likely to rely heavily on coal, will be putting out a far higher proportion of the world's CO$_2$ in 2025 than they are today. (Note: 'US etc.' includes Canada; 'W. Europe etc.' includes Japan, Australia, New Zealand; 'USSR etc.' includes Eastern Europe; 'China etc.' includes N. Korea and Vietnam. Source: R. M. Rotty in Carbon Dioxide, Climate and Society: IIASA *Workshop Proceedings, 1978, ed. J. Williams, Pergamon, Oxford, 1978.)*

China etc.
8%

Third World
13%

US etc.
29%

USSR etc.
25%

W. Europe etc.
25%

1974

US etc.
9%

China etc.
16%

W. Europe etc.
11%

Third World
33%

USSR etc.
31%

2025

South, more concerned with the immediate benefits of industrial growth than with any longer-term hazard to the global climate. This makes the carbon dioxide problem a global political issue, and as we shall see in Chapter 11 the political implications are as complex as the scientific problems. The simplest of these new issues has already been seized on by the nuclear lobby. If the threat of carbon dioxide pollution is seen as real, and if the poor South is going to increase greatly its use of coal, then there may be an urgent need for the USA and Europe not merely to restrain growth in consumption of fossil fuel but actually to cut back its use severely. The real world is much more complicated, but such arguments are already being used by supporters of the nuclear option to press the case for a major shift towards nuclear power as the only easy route away from fossil fuel. Alternatively, as Rotty puts it, 'the problem of avoiding CO_2 triggered climatic change becomes that of providing fuel for the developing countries to assure their progress without such heavy dependence on fossil fuels' – an equally interesting political problem.

Since 1978 Rotty has produced some revised scenario calculations which take account of the recent lower rates of growth in global energy demand. Following a meeting in Stockholm to discuss the carbon dioxide problem early in 1981, Professor Bert Bolin of the University of Stockholm told me that at a World Meteorological Organization meeting late in 1980 Rotty had presented a revised scenario based on a build-up of carbon dioxide in the atmosphere to 450 ppm in 2025. The original 'business as usual' scenario implied an output of 23 Gt of carbon in the form of carbon dioxide in 2025 alone, producing an increase in atmospheric carbon dioxide of 11 ppm in the course of a single year, assuming that half stays in the atmosphere. The revised, lower figures correspond to burning between 12 and 15 Gt of carbon in that year, still an input of 8 ppm of carbon dioxide to the atmosphere.

Given all the uncertainties of energy forecasting, these figures are as reliable a guide as any. Unless deliberate efforts are made to avoid releasing carbon dioxide to the atmosphere, the pre-industrial concentration of 300 ppm could double, to 600 ppm, by 2025, and will very likely increase by 50 per cent, to 450 ppm. The question the climatologists have to answer is what effect such a considerable change would have on the weather of the world.

Climate in a computer

The first computer models of world climate that were able to make sensible forecasts of the effects on world weather of an increase in the atmospheric concentration of carbon dioxide were being developed in the early to mid 1970s, at a time when to many people the 4-percent annual growth ('business as usual') scenario still looked like a good approximation to the real world, and some people still talked of economic growth, and a corresponding increase in energy demand, rapid enough to cause a doubling of the atmospheric carbon dioxide concentration even before the end of this century. In that climate of opinion, those early computer models of the likely effects of the build-up on climate came as something of a scientific bombshell.

Because of the natural variability of climate – the 'background noise' in the system, as the computer modellers call it – small changes in rainfall, temperature and so on are difficult to identify and interpret. If we have one hot summer – like that in the USA and Canada in 1980 – it doesn't necessarily mean that the world is getting hotter, through the greenhouse effect or for any other reason, but is most likely just one of those things that happen by chance. At the same time, even today the computer models developed by climatologists are not very good at finicky, detailed work, but are most reliable when used to provide broad outlines for the big picture, painted with a big brush. The combination of these two effects means that throughout the 1970s, and to a large extent even today, climatologists have been extremely cautious about taking at face value any computer forecasts of the effect on climate of anything less than a doubling of the pre-industrial concentration of carbon dioxide.

Strictly speaking, the computer models do not tell us anything about the effects of lesser increases, but few people can resist making the seemingly reasonable guess that a smaller increase (say, by 50 per cent) will produce a proportionately smaller change (in this case, we might guess, half as much), but in the same direction. That is only a guess, and may not prove so reasonable if it turns out that some changes only happen when a certain threshold level of carbon dioxide is reached. The computer models may give a good guide to what will happen when the carbon dioxide concentration has doubled – though some of the experts doubt this – and they may, therefore, indicate the direction the weather machine must be moving in now. But they do not provide a guide to the route the weather machine will follow to reach the next landmark along the way.

Until the mid 1970s the computer models developed by different groups with their forecasts for a world with an atmosphere twice as rich in carbon dioxide didn't even agree with one another. These early models produced a variety of estimates of the effects of such a carbon dioxide doubling on temperature, and were virtually unable to say anything at all about how rainfall patterns might change in a warmer world. The situation began to improve in 1975, as the models themselves were improved.

The simplest models used a global average in which the heat input from the Sun was balanced against the heat radiated by the Earth, taking account of infra-red absorption by carbon dioxide, but using only one value for the average temperature of the globe at each step in the calculation. The next step was to introduce an allowance for the difference in temperature at different latitudes, using average temperatures for different latitude zones. Some of these 'two-dimensional' models also attempted to allow for feedback from cloudiness, the movement of energy from the tropics to the poles, and so on. But the models now most widely favoured in climate studies are the 'three-dimensional' General Circulation Models, or GCMs. These models start out with an initial set of conditions (solar heat, cloud cover, zonal distribution of temperature both with latitude and, in the best models, height through the atmosphere, carbon dioxide concentration, and so on). They then 'run', performing repeated calculations of the way the atmosphere responds to these conditions, and how the parameters respond to the changes in the atmosphere, until they arrive at a steady state with all the variables constant. These values of the key parameters are then taken as a guide to how the real world would respond to similar starting conditions.

Even the three-dimensional GCMs are far from accurate representations of reality. In most GCM calculations only one hemisphere of the globe is considered, and the solar radiation has to be set at a fixed value corresponding either to summer or winter. The models cannot calculate the effects of seasonal variations. Since such complex models have to be run in the computer for a long time to produce results, and many different runs are needed to provide a statistical basis for interpreting their 'forecasts', they are extremely expensive in scientific research terms; only a few have been developed, and these were produced by the teams with access to the biggest and fastest computer systems.

Perhaps most serious of all, no GCM yet developed deals adequately with the oceans. Sea surface temperatures are set in the models but usually do not change in feedback with the other parameters, and the role of ocean currents in carrying heat around the world is ignored. This is a very crude approximation to reality for a planet whose surface is 70 per cent covered by water. One group, at the US Geophysical Fluid Dynamics Laboratory in Princeton, has attempted to include ocean effects using a computer model in which the sea is represented by a non-circulating swamp, which has no capacity to absorb heat and has an infinite supply of water; this is scarcely less crude. The 'best' models, according to Professor Bolin, are now beginning to use numbers corresponding to an ocean in which the surface layer is mixed down to a depth of 100–200 metres. But this, he says, is 'still inadequate'.

The GCMs are the best tools available for studying different climatic regimes, but they work best when looking at the big picture – the differences in atmospheric circulation between the present day and a full Ice Age, for example. As used in the greenhouse-effect scenarios, they are operating on the limit of their useful reliability. Climatologists recognize that the models are imperfect and that their 'predictions' should not be taken as the last word; but science is all about probabilities, and a great many climatologists are now convinced that the probability that the GCMs are broadly correct is too great to be ignored by governments and others involved in long range economic planning.

The 2°C consensus

Agreement between the different computer modelling teams began to emerge in 1975. In what is now seen as a key paper in the development of the debate about effects of increased carbon dioxide concentration on the climate, Stephen Schneider of the US National Center for Atmospheric Research in Boulder, Colorado, attempted to clear up the apparent confusion caused by the wide range of estimates which had been published by different teams. The published figures for a doubling of the atmospheric concentration of carbon dioxide ranged from a low estimate of an increase in temperature of 0·7° C to the highest estimate of 9·6° C. Schneider examined the different assumptions made by the climatologists using each computer model and related them to the physical conditions in the real world. He found explanations for most of the differences between the published forecasts and concluded that the best estimate for the temperature increase corresponding to a doubling of the atmospheric carbon dioxide concentration was a rise of between 1·5° C and 2·4° C.

In the same year Syukuro Manabe and Richard Wetherald of Princeton University published what was then one of the most complete GCM simulations ever carried out, using the Geophysical Fluid Dynamics Laboratory model with its ocean swamp. Their figure for the temperature increase due to a doubling of carbon dioxide also came out close to 2° C, nicely in the middle of the range spanned by Schneider's interpretation of all the available modelling evidence.

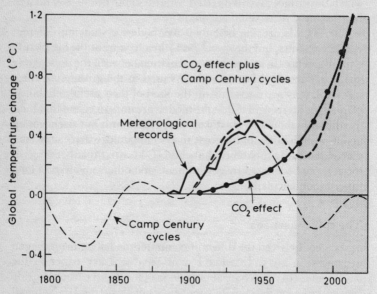

Figure 9.4. Wallace Broecker's combination of global temperature changes due to natural cycles with the predicted carbon dioxide greenhouse effect, and the resulting temperature forecast. This was the first scientific forecast in recent times to draw attention to the possibility of a rapid global warming in the near future. (Source: *W. S. Broecker, 'Climatic change: Are we on the brink of a pronounced global warming?', Science, vol. 189 (1975), p. 461.*)

The combination of Schneider's work and the GCM calculations by the Princeton team would probably have been enough to ensure that the rule of thumb that a doubling of carbon dioxide corresponds roughly to a 2° C rise in global average temperatures became enshrined in the hearts of many climatologists. But a third scientific paper also published in 1975 brought this figure – really still no more than a rule of thumb approximation – to the attention of a much wider audience, and helped to trigger the more widespread debate which has continued – and grown – until the present day.

Wallace Broecker of the Lamont-Doherty Geological Observatory was one of the first respected scientists to look in detail at the implications of a carbon-dioxide-induced global warming. As a basis for his prognostications he used a value of 2·4° C for the global warming due to a doubling of carbon dioxide, a number which came from some older calculations made by Manabe and Wetherald in 1967, but which turned out to be close to the consensus figure. Either Broecker was lucky, or he was an astute judge of the reliability of the different models. He certainly picked a good place to publish his discussion of the seemingly imminent global warming – the pages of *Science*, probably the world's most widely read scientific journal, and one from which the popular media often pick up exciting scientific stories. Using a combination of the natural temperature cycles found from isotope analysis of the Greenland ice cores, some simple 'business as usual' assumptions about the growth in atmospheric carbon dioxide, and an estimate that each 10-per-cent rise in carbon dioxide concentration brings an increase in temperature of 0·3° C,* Broecker caused a minor sensation with his forecast that 'the exponential rise in the atmospheric carbon dioxide content will tend to become a significant factor and by early in the next century will have driven the mean planetary temperature beyond the limits experienced during the last 1000 years'. Since the ice-core cycles suggest that a natural cooling trend is about to reverse, just as the anthropogenic

* The figure of 0·3°C temperature rise for a 10-per-cent increase in carbon dioxide concentration comes from the estimate that temperature increases in proportion to the logarithm of the atmospheric carbon dioxide concentration. The basis for this depends on assumptions about the exact way infrared radiation is absorbed as the carbon dioxide concentration builds up; it may not be entirely accurate, but it helped Broecker to make some detailed and dramatic forecasts that caught the attention of the scientifically aware public.

greenhouse effect may be becoming significant, Broecker suggested that 'the onset of the era of CO_2-induced warming may be much more dramatic than in the absence of natural climatic variations ... we may be in for a climatic surprise'.

A detail Broecker mentioned only in passing, almost as an afterthought, may have been one of the most important points in his article, as we shall see. 'Manabe and Wetherald have suggested', he mentioned, 'that the effect in polar regions is much larger.' The computer models of the late 1970s, and other studies, have shown that this is indeed the case. A global *average* warming of a degree or two may bring with it an increase in temperature several times larger at high latitudes, but only a small increase in the tropics. The reasons for this are connected with a more efficient transport of warm tropical air polewards in a warmer world, and with the way snow and ice cover reflects away incoming solar heat. Any small temperature increase which reduces the extent of snowfields and ice sheets means that there is a larger area of dark ground or sea surface to absorb solar energy, ensuring a further increase in surface temperatures. In the tropics, all the heat that can be absorbed already is being absorbed, so this feedback effect has no chance to get going.

The implications of the regional effects of even a modest increase in the carbon dioxide concentration of the atmosphere are all-important. New model calculations, and other ways of investigating the regional patterns of climatic change associated with a global warming, have, in the early 1980s, provided a very different perspective on the true nature of the carbon dioxide problem. In a nutshell, the regions which will be most dramatically and most quickly affected by the changes caused by the anthropogenic greenhouse effect are the regions at high latitudes – the rich North. But, as I have already mentioned, the regions doing most to cause the build-up of carbon dioxide in the immediate future are those at lower latitudes, the poor South. Whether or not the rich North is concerned about the carbon dioxide greenhouse effect, is the poor South, which may be largely unaffected and might even benefit from the expected climatic changes, going to care two hoots?

Most of the rest of this book is concerned with the eye-opening implications of this new, and still largely unpublicized, aspect of the carbon dioxide problem. But before moving on to the regional implications of a global warming – for rainfall, temperature and agriculture – it is only right to acknowledge that there is a minority view which holds that the 2° C consensus is wrong. The consensus comes, in fact, from only a handful of groups – literally three or four teams – working with the big GCM models. Could it be that they are all making the same mistake somewhere in their calculations, and are collectively barking up the wrong tree?

A voice of dissent

Sherwood Idso of the US Water Conservation Laboratory in Phoenix, Arizona, set the cat among the climatological pigeons when he published a short paper in *Science* early in 1980, suggesting that all the climate modellers were making the same mistake, and that the 2° C consensus was roughly 10 times too big. According to Idso, a doubling of the atmospheric concentration of carbon dioxide would cause a global warming of no more than a quarter of a degree centigrade. The modellers were understandably upset by this suggestion, and they were especially upset because Idso's brief *Science* paper gave no details of the method he used to arrive at this conclusion, instead referring interested readers to a longer paper 'in preparation'. This is a normal procedure with items of hot scientific news, not so much to give the researcher two bites at the publicity cherry as to ensure that the important conclusions reach a wide audience quickly, while the complex and, let's face it, sometimes boring details are tucked away in weighty, slow-publishing journals. It means there is a pause before the details are readily available, but interested parties can always get these direct from the researcher concerned – in this case, Idso – and Idso has been at pains to point out that all the relevant background material was seen by *Science* before they accepted his shorter paper. Happily, Idso was able to present his full argument in detail to a 'workshop' on 'Responsible Interpretation of Atmospheric Models and Related Data' held at the Scripps Institution of Oceanography in March 1981, and his detailed paper should have been published in the book of the workshop proceedings (from the American Institute of Physics) by the time you read this. The modellers may remain unconvinced by Idso's arguments, but at least they are now there in black and white for all to see.

The arguments are especially interesting because Idso has used a completely different conceptual approach to the carbon dioxide problem. Instead of trying to simulate the weather machine in a computer and calculate how it will respond to a build-up of carbon dioxide, he has looked at the way temperature at the surface of the Earth varies in the real world when atmospheric conditions change, and tried from these observations – direct measurements of temperature and radiant energy – to calculate a 'response function' which will tell us how the temperature will respond to the carbon dioxide greenhouse.

During his studies of the balance between incoming solar energy and outgoing infra-red from the ground in Arizona, Idso has monitored changes produced by dust in the atmosphere, variations in moisture content (humidity) and the changing cloud cover. Dust storms, for example, make very little difference to ground temperature in daylight hours, because the heat they block from the Sun is compensated for by the heat the dusty atmosphere radiates back to the ground. Even more interesting, there is a time in late June or early July each year when the amount of water vapour in the air over Arizona increases dramatically, doubling or tripling almost overnight. This Arizona monsoon is a result of high-level winds from the Gulf of Mexico and low-level moisture 'surges' from the Gulf of California. Using records going back over the past 30 years, Idso found that an increase in the moisture content of the atmosphere sufficient to raise the surface vapour pressure of water from 4 to 20 millibars corresponds to an increase in the surface air temperature at dawn of $11 \cdot 4°$ C. This large increase in the minimum temperature of the daily cycle – it may or may not be darkest before dawn, but it is certainly coldest, at least in the desert – shows the blanketing effect of the water vapour to best advantage. The daily maximum temperatures, and averages, show only a very much smaller effect, for the same reason that a blanket of dust has little effect on daytime temperatures – the heat blocked from the Sun on the way in almost balances the heat from the ground blocked on the way out. From dust and 'monsoon' studies together, Idso calculated that for each extra watt of energy passing through each square metre of air just above the ground, the surface air temperature responds by an increase of $0 \cdot 196°$ C.

Both these calculations, however, depend on changes in the atmosphere over the span of only a few days. To compare them with seasonal changes, Idso used a compilation of measurements of solar radiation from 105 observing stations scattered across the USA, and compared them with seasonal temperature fluctuations at the observing sites. He found that inland sites showed just the same forcing function – 0·19° C per watt per square metre – but that for sites on the edge of the ocean the response was halved. Assuming that this represents the biggest possible response of the oceans, and allowing for the fact that 70 per cent of our planet is covered by sea, Idso estimated that the global average response function must be no more than 0·113° C per watt per square metre.

What does this tell us about the greenhouse effect? It is a relatively simple matter to calculate the increase in the amount of radiant energy at the ground which would occur if the carbon dioxide concentration increased from 300 ppm to 600 ppm. It is 2·28 watts per square metre. Multiplying this by the forcing factor 0·113 gives an equivalent rise in global mean temperatures of 0·25° C, in stark disagreement with the computer models.

Some similar calculations by Reginald Newell of MIT and Thomas Dopplick of Scott Air Force Base produced a similar figure for the forcing function, which Idso sees as confirmation of the validity of his technique. The critics, among them Stephen Schneider, Will Kellogg and V. Ramanathan of the US National Center for Atmospheric Research, did not agree, and said so in a letter to *Science*. Their main argument was that Newell and Dopplick's approach – and Idso's too – ignored a crucial feedback mechanism. In the real world, extra warmth causes more water vapour to evaporate from the oceans, which increases the moisture content of the atmosphere and, because water vapour is a good greenhouse gas, increases the temperature still more. 'We are not persuaded that Idso's results should alter our perception [of the carbon dioxide problem],' said the NCAR team. In his Scripps presentation Idso attempted to answer that major point of criticism. For an air temperature of 15° C, he said, the increase in water vapour pressure produced by a 0·25° C warming is just 0·2 millibars. This in turn produces a sufficient additional greenhouse effect to increase the temperature only by a further 0·07° C, and that temperature increase in its turn, sending a little more moisture into the air, causes a further increase of only about 0·01° C in global mean temperatures. We have rapidly reached the point of diminishing returns, and even with this extra effect the calculation still comes up with an increase of only about 0·3° C for a doubling of the carbon dioxide concentration.

A physicist myself, I have great sympathy with Idso's approach to the carbon dioxide problem, using real measurements of radiation and temperature changes in the real world as the basis for his calculations, rather than abstract numbers locked away in the bowels of a computer. One observation is worth a ton of theory. But even as a physicist I find the details of the radiation balance measurements too specialized to be familiar. And, of course, even the seasonal measurements are dealing only with annual rhythms, whereas the carbon dioxide build-up is a problem on a time-scale of decades and centuries. James Hansen (whom we met in connection with the studies of the climatic effects of the Mount Agung volcanic eruption) and his colleagues at the NASA/Goddard Space Flight Centre have raised just this point, arguing in a paper published in *Science* in August 1981 that it takes years for the oceans to respond to a long-term change in the energy balance of the globe. Their computer model, when set the task of calculating a response function like the one used by Idso, came up with a figure of $0.2°$ C per watt per square metre inland, half that on the coast, and only one tenth as much over the ocean, broadly agreeing with Idso's measurements in the real world. But the immediate result of an instantaneous doubling of carbon dioxide concentration in the computer model is that most of the extra energy goes to warm up the oceans. The *eventual* increase of surface temperature (in this particular model, by $2.8°$ C) occurs only after the oceans, the greatest reservoirs of heat and sources of 'thermal inertia', have adjusted to the new balance between incoming and outgoing radiation. The $2°$ C consensus applies only to conditions when a new equilibrium has been established. As long as the carbon dioxide concentration is changing, of course, no new equilibrium can be established, and if nothing else Idso's work highlights the fact that the computer models tell us little or nothing about how the world will get from its present climatic state to the predicted pattern appropriate for a world with twice as much atmospheric carbon dioxide.

Idso still has a shot in his locker, though, and in many ways it is his most impressive. Forget all the details, he argues, and look at the Earth as a whole. Manabe and Wetherald's expertise in computer modelling was also used, back in 1967, to assess the equilibrium temperature of a world just like the Earth today but with no atmosphere. This rather unlikely world would still have oceans, vegetation, mountains, desert and so on, but no blanket of air around it and so no greenhouse effect at all. Its temperature, said Manabe and Wetherald, would be $-23°$ C (as Idso puts it, the 'global air temperature' of that airless globe would be $-23°$ C!). A more recent improvement to this calculation suggests that a better value would be about $-19°$ C, and, since the present global mean air temperature is close to $15°$ C, the implication is that the overall greenhouse effect of the atmosphere is to increase temperature by some $34°$ C.

This includes every feedback process involving oceans, water vapour, clouds, dust and anything else you can think of, plus everything you don't think of, and it represents an equilibrium that has been established over thousands of millions of years. The 'perturbation' responsible is the absorption of some of the outgoing radiation from the Earth's surface by the atmosphere, and that total outward flux is measured as 348 watts per square metre. So the appropriate response function for the whole Earth over geological time-scales is $0·1°$ C per watt per square metre, simply from dividing the $34°$ C of the greenhouse effect by the measured flux of energy. Once again, the figure is close to the value found by the local studies. But the computer models, which show a much bigger carbon dioxide greenhouse effect than Idso's calculations, also include within themselves this sort of value of the response function.

Idso's next step is to look at a property of the atmosphere called its emissivity, which is a measure of how closely its properties as an absorber and radiator of energy compare with the most efficient radiator possible, which is termed a 'blackbody'. A blackbody is in this sense really an abstraction, a physicist's idealized view of reality. But many objects – the Sun, for example – behave very nearly like blackbodies in many ways. A perfect blackbody absorbs all the energy which falls upon it, and for the outgoing radiation from the Earth's surface (*not*, of course, for the incoming solar energy at much shorter wavelengths) the atmosphere of the Earth is about 90 per cent as efficient as a blackbody at absorbing energy. *Whatever* happened to the atmosphere, it could never be more efficient than a blackbody, and acting now as a 90-per-cent blackbody in the infrared it has produced a greenhouse effect of less than $40°$ C overall. That extra 10 per cent of efficiency, says Idso, could not possibly produce more than another 10-per-cent greenhouse effect – no more than a further rise in global mean temperatures of $4°$ C.

This somewhat startling – to anyone familiar with the GCM forecasts – conclusion applies to the carbon dioxide greenhouse effect and any other change in the atmosphere, short of some drastic physical change like increasing its density to match that of the Venus atmosphere. Whatever the carbon dioxide may do, it can't make the atmosphere more efficient at its greenhouse-effect warming than a blackbody would be, and, according to Idso's calculations, that fixes an upper limit on the temperature of the Earth today only $4°C$ above present mean temperature. Geological evidence supports Idso's case. For at least millions of years, and almost certainly for very much longer, the Earth has never been more than a couple of degrees warmer than it is today, even though the amount of carbon dioxide and other trace gases in the atmosphere may have changed considerably.

The real world seems, on this evidence, to be less sensitive to temperature variations than the GCMs would have us believe, and even Wallace Broecker, the man who first really rang the alarm bells about the carbon dioxide build-up, can be found unwittingly providing support for Idso's case. At the Carbon Dioxide and Climate Research Program Conference held in Washington, DC, in April 1980, Broecker was asked whether it might not be possible to find out what a warmer world would be like by reconstructing, from geological evidence, the climate of some recent past epoch when the world was $2\frac{1}{2}°C$ warmer than it is today. He replied that 'there may never have been a time that warm, at least in the last few million years'; and another contributor to the discussion added that 'past climates differ from today's climate more in terms of spatial patterns of temperature and atmospheric circulation than in terms of global mean temperature'.

To a physicist with no professional reputation at stake however the argument is resolved, this last example from Idso commands respect and suggests that at least we shouldn't get too excited about the GCM consensus. In fairness, the computer modellers never said we should, and have pointed out all along the uncertainties in their work. These uncertainties have, however, become very much clearer as a result of Idso, Newell and Dopplick setting the cat among the pigeons. By October 1981, other researchers had begun to look at the whole problem of forecasting the greenhouse effect afresh, and Robert Kandel of the Pierre and Marie Curie University in Paris had made the same kind of comparison of the new approach and the latest GCM calculations as Schneider made for the various computer models back in 1975. In an article published in *Nature*, Kandel pointed out that the apparent disagreement between the two approaches to the problem resulted from one real difference, the way they tackled the question of evaporation from the oceans. Both because it takes a lot of heat energy to evaporate water, and because extra water vapour makes its own contribution to the greenhouse effect, this is a key question, but one which is seldom handled in a satisfactory manner, according to Kandel. The exact value of the temperature increase forecast for a doubling of carbon dioxide in the atmosphere depends, in all the calculations, on the difference between two relatively large numbers which are almost in balance and describe the income and expenditure of the surface energy budget. One of those larger numbers itself depends on the assumptions made about humidity changes in the air just above the sea surface, so a small change in the humidity parameter leads to a big change in the forecast temperature increase.

Making a plea for computer modellers to publish more information about the humidity assumptions inherent in their models, so that other people have a better chance of understanding their forecasts, Kandel says that, as far as we can tell today, 'depending on the response of the hydrological cycle and of atmospheric humidity to a doubling of CO_2, the rise in surface temperature could be as low as 0·5 K, or as high as 10 K'. Here the degrees K are exactly equivalent to degrees C, and it looks as if we are back, almost exactly, to the confusion of 1974, before the original computer consensus emerged. In view, however, of Idso's calculation of the total greenhouse contribution of the whole atmosphere at present, and the suggestion that the largest possible further increase in greenhouse temperature of the Earth is no more than 4° C, it seems most likely that the bottom half of the range Kandel mentions, from 0·5° C to 5° C, can safely be taken as encompassing the possible increase of global temperature if carbon dioxide concentration doubles. This still leaves more uncertainty than most forecasters seem to credit, together with the suspicion that there may not be much of a problem there at all. But the case is not yet proven either way, and even if Idso's voice of dissent eventually turns out to be telling us more about the real world in this particular regard than the GCMs do, that mention of spatial patterns pulls the rug from under any growing feeling of complacency.

Spatial patterns describe the regional distribution of wind, rain and temperature. If the high latitudes are affected much more than the tropics, a warming of even 1° C (which might satisfy the calculations of both Idso and the GCMs) could bring a change of several degrees in key regions of the globe. And, as a warming trend brings changes in other climatic factors, it might be that new patterns of drought and flood become important to agriculture even before the globe warms by one degree. The concern about regional changes of climate in a warmer world, which began to build up in the first two years of the present decade, is scarcely affected even by taking all of Idso's calculations at face value. In addition, there is already some evidence that although the expected global warming is not happening as fast as the GCMs have predicted, it may already be measurable, and at a level greater than Idso's calculations would imply. The truth may lie somewhere between the two, and there is still ample cause for concern.

10

Carbon Dioxide and Man: Adjusting to Change

In their excellent book *Thinking Physics*, Lewis Epstein and Paul Hewitt (Physics teachers at the City College of San Francisco) caution their readers: 'don't rely on words, or equations, until you can picture the idea they represent'. This aphorism ought to be engraved on the hearts of all physical scientists. If you look only at the graphs and equations, and you see an exponentially rising build-up of carbon dioxide in the atmosphere coupled with numbers showing that carbon dioxide traps infra-red radiation, you might easily be frightened into imagining the world set on course for a runaway greenhouse effect, doomed to fry like the planet Venus. But there is a limit to how much radiation is available to be absorbed. The Earth only radiates so much infra-red energy, and no matter what the rise in carbon dioxide concentration may be, there has to be a limit to the rise in temperature.

At some concentration of carbon dioxide in the atmosphere, all of the radiation emitted from the Earth's surface in the 12 to 18 micron infra-red band would be being absorbed. Once this state was reached, the addition of more carbon dioxide to the atmosphere could not warm the planet further, because no more outgoing radiation of the right wavelength would be available. Adding all of the atmosphere of the Earth to a bare planet has produced a rise in temperature of only 34° C, yet measurements show that in the crucial infra-red radiation bands the atmosphere is already absorbing nearly 90 per cent of the outgoing radiation. So, Idso argues, filling in the infra-red band completely – in blackbody terms – cannot cause a further rise in global mean temperatures of more than 4° C.

But how quickly could the temperature increase, whatever the limit may be? The carbon dioxide infra-red absorption band overlaps with the absorption band of water vapour, and there is already a great deal of water vapour in the atmosphere. Infra-red radiation that is already absorbed by water vapour cannot be absorbed again by carbon dioxide, so the 'window' which carbon dioxide might fill in is small. Any specific increase in carbon dioxide concentration today – a doubling, say – will cause only half the rise in temperature that it would if there were no water vapour in the atmosphere. The 'increase' corresponding to the other half of the carbon dioxide increase has already happened, thanks to the water vapour, and is part of the 34° C rise due to the whole atmosphere.

Newell and Dopplick, whose independent calculations produce the same sort of numbers as Idso's calculations of the likely carbon dioxide greenhouse effect, emphasize that water vapour, not carbon dioxide, actually dominates atmospheric absorption of outgoing heat radiation, and they suggest that the high figures for the greenhouse effect produced by the GCMs are due to a mistake. The models assume that evaporation from the oceans in the tropics increases as temperature rises. This assumption has never been checked or tested in practice. Even if it is correct, Newell and Dopplick argue that changes in cloud cover must be included in the models before any credence can be given to the GCMs. Increased evaporation would produce more clouds, more clouds would reflect away more of the incoming solar radiation, and this would counterbalance the influence of an enhanced greenhouse effect. The heat movements 'controlled' by carbon dioxide are only 10 per cent of those 'controlled' by water vapour, say Newell and Dopplick, arriving at the same conclusion as Idso – that the atmosphere is already a 90-per-cent perfect infra-red greenhouse – by a different route.

None of this is really new. The climate modellers all acknowledge the role of water vapour and cloud cover, and they all agree among themselves on the deficiencies of the GCMs. But even so the carbon dioxide warming 'predicted' by the GCMs is often reported as a hard and fast number, so it is important to spell out the deficiencies in the models here. The argument will only be resolved when the expected global warming becomes big enough to be detected and measured. For there are two sides to the coin: whatever the warming, there must be a limit; but whatever the limit, there must be *some* warming. The evidence so far is not on the side of the GCMs, for according to some theorists the increase in temperature forecast by the three-dimensional climate models should already be detectable now, and it is not. Ironically, one of the people who has recently pointed this out (writing in *Science* with Roland Madden) is V.

Ramanathan of NCAR, who was one of the first climatologists to criticize Idso's suggestion that all the GCMs might be making the same mistake.

Where is the expected increase?

Madden and Ramanathan commented on the inherent variability of climate which makes detection of changes due to a build-up of carbon dioxide difficult – something that should be very clear from Part One of this book. But they went on to consider climate changes caused by carbon dioxide as a 'signal' compared with the 'background noise' of natural variability, and attempted to provide estimates of how big the signal ought to be, compared with the noise, and whether or not it should yet be detectable.

Using monthly average temperature records from sites around the globe close to latitude $60°$ N covering the period from 1906 to 1977, they were able to get a physical picture of how much natural variability of temperature there has been during the present century. Over the same time-span the observed increase in atmospheric carbon dioxide concentration, combined with the computer models, predicts that the whole pattern of variability should have been bodily shifted to higher temperatures by a measurable amount. When they ran all the statistics through their computer, Madden and Ramanathan found that for these observing sites the pattern of '20-year average temperature from 1956 through 1975 is not higher than a 20-year average temperature from 1906 through 1925; in fact, it is slightly lower ($14·2°$ C compared to $14·42°$ C). Therefore we cannot provide statistical evidence that there has been an effect due to increasing CO_2 on the present mean zonal temperature at $60°$ N.' And they concluded that 'either such models [the GCMs] overpredict the signal, or other compensatory climate changes are occurring'.

Reid Bryson's 'human volcano' is the most obvious candidate for some other 'compensatory climate change'. Bryson and his colleague G. J. Dittberner, at the University of Wisconsin–Madison, have developed a model of climatic changes in the present century which includes effects of both dust and carbon dioxide, and from this they conclude that dust accounts for as much as 90 per cent of the temperature variation in recent decades, while carbon dioxide accounts for only about 3 per cent. This is very much a minority viewpoint, but from the evidence of studies like that of Madden and Ramanathan such a possibility has to remain in the arena of scientific debate about current climatic changes.

Undaunted by the lack of a detectable carbon dioxide signal in temperature records so far, Madden and Ramanathan point out that it will be easiest to detect the effects, when they do become measurable, in records of summer temperatures. If there is no cooling effect acting to cancel out the expected warming, then it should become detectable any time between now and the end of the present century – sooner if the GCMs are nearly right, later if they are anything like as wrong as Idso believes.

Madden and Ramanathan's study appeared in August 1980 and took the seemingly desirable approach of using as long a run of monthly temperature measurements as possible, comparing temperatures in the past two decades with those early in the twentieth century. In August 1981 another paper from Hansen's team at the NASA/Goddard Space Flight Center appeared, looking at the same problem but highlighting temperature changes between the middle 1960s and 1980, as well as looking at the broader picture. The temperature of the Northern Hemisphere decreased by about $0.5°$ C between 1940 and 1970, which is why Madden and Ramanathan found no detectable greenhouse effect in their study, even though this was a time of rapidly increasing carbon dioxide concentration. But although this global cooling is now part of climatological folklore, the Goddard team points out that the latest temperature measurements, broken down by geographical regions, reveal a sharp turnaround in the temperature curves in the past decade or so.

To put these changes in the perspective of the past hundred years, the high northern latitudes ($23.6°$ N to the pole) warmed by $0.8°$ C between the 1880s and 1940, then cooled by about half a degree between 1940 and 1970. The low latitudes in both hemispheres warmed by about $0.3°$ C between 1880 and 1930 and have not changed very much since then. The high southern latitudes ($23.6°$ S to the pole) warmed by about $0.4°$ C over the past century, in a more steady manner than the high northern latitudes. Putting all the data together (see Figure 10.1), Hansen's team concludes that 'the global temperature is almost as high today as it was in 1940. The common misconception that the world is cooling is based on Northern Hemisphere experience to 1970.' Since 1970 the Northern Hemisphere has once again begun to warm up.

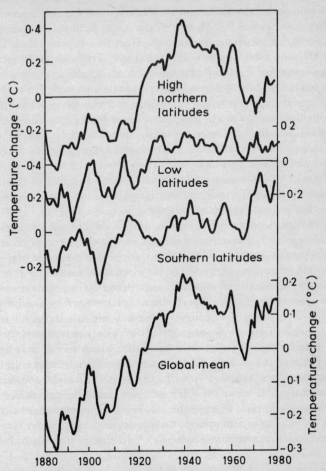

Figure 10.1. The best estimates of temperature changes in different latitude bands, and for the globe as a whole, for the past hundred years. Since 1970, the world has begun to warm up. (Source: *J. Hansen et al., 'Climatic import of increasing atmospheric carbon dioxide', Science, vol. 213 (1981), p. 961.*

How does this warming – 0·4° C worldwide since the 1880s – compare with the computer models of the greenhouse effect? The Goddard group has used a one-dimensional computer model in their study. This ignores latitudinal variations and calculates temperature through the atmosphere for an averaged-out globe as a function of altitude. Its advantage is that although it is less detailed than a full three-dimensional GCM, it provides scope to mimic in the computer the effect of changing various other factors at the same time as allowing for increased carbon dioxide. The very detail which ought to make the three-dimensional models more accurate leaves no room in the computer to look at other factors, as well as making the models expensive to run. So Hansen and his colleagues have used a one-dimensional model which they know gives similar forecasts to the GCMs when used with carbon dioxide as the only variable, and have compared forecasts of the carbon dioxide greenhouse effect with and without feedback from water vapour, with and without the effects of volcanic dust, and with or without estimates of the (admittedly speculative) effect of a variation in the amount of heat from the Sun getting through to the troposphere.

Such calculations are, of course, only as good as the assumptions about each of those factors fed into the computer. The programmers' aphorism is 'garbage in, garbage out', which means if you put rubbish into a computer don't be surprised if the numbers that come out are useless. In this case, the assumptions about solar variations are probably a weak link – news about the latest work by Kondratyev and Nikolsky had not made its way across the Atlantic at the time the Goddard study was being carried out. The preferred model from this study has a fixed relative humidity, which means that more water vapour builds up in the atmosphere as the temperature increases – the assumption Newell and Dopplick, Kandel and others question – and leaves the solar influence out of the calculation. It produces a forecast of a rise in global mean temperatures of $1 \cdot 9°$ C for a doubling of atmospheric carbon dioxide, right on the nose of the computer modellers' consensus.* In this model, high clouds

---

* With fixed *absolute* humidity, so that there is no feedback effect from increasing water vapour, and with everything else except carbon dioxide concentration held steady, this model comes up with a forecast global warming of $1 \cdot 2°$ C for a doubling of the amount of carbon dioxide in the atmosphere. This is still very much bigger than the size of the warming predicted by Idso using a different approach, or by Newell and Dopplick using their computer model with no water vapour feedback. If we were talking in abstract scientific terms – speculating about the temperature on some distant planet, perhaps – we would reckon that all these numbers were in the same ball-park and the scientists would pat one another on the back in delight at their agreement. It's only because the planet being studied is

contribute a net greenhouse effect, while low clouds cool the surface. But like all computer models to date, it cannot approximate the detailed effects of cloud cover accurately.

Hansen has also been involved in modelling the effect of the Mount Agung eruption in the atmosphere, so his group has a sound basis for including the effects of volcanic dust in their calculations (although Kondratyev and Nikolsky would argue, perhaps, that they are allowing twice as much influence from this source as they should, since half the cooling in the early 1960s attributed to Agung was, the Russians believe, due to the influence of nuclear bomb tests). And a Goddard team headed by Wei-Chyung Wang has been involved in detailed calculations of the possible influence of a greenhouse effect from trace gases such as methane and oxides of nitrogen, so when the Goddard group says that the net effect of anthropogenic trace gases to date ought to be a warming of about $0.1°$ C, which 'does not greatly alter analyses of the temperature change over the past century', they ought to know what they are talking about there too.* Putting everything together, they come up with a combined carbon dioxide plus volcanic dust influence which looks intriguingly close to the pattern of temperature changes in the real world over the past hundred years (Figure 10.2). But it is very clear from this illustration that other factors at least as important as the supposed greenhouse effect (perhaps the solar influence?) have been at work in the past century, and that this study, like that of Madden and Ramanathan, does not provide definitive proof that the greenhouse effect is at work.

Projecting these trends on the assumption that their computer model is a good guide to the greenhouse effect, Hansen and his

our home, and temperature differences of a degree or so may be important to agriculture, that we complain about the disagreement and the two camps seem to be at odds with one another. Give them time and they'll sort out their differences and reach even closer agreement. But how much time have we got?

---

* When they say, however, that trace gases 'will significantly enhance future greenhouse warming if recent growth rates are maintained', perhaps the comment should be taken with a pinch of salt. Remember there is a limit to how much infra-red radiation is around to be absorbed; once the 'window' in the greenhouse is blocked up, adding more panes of glass on top won't have much effect.

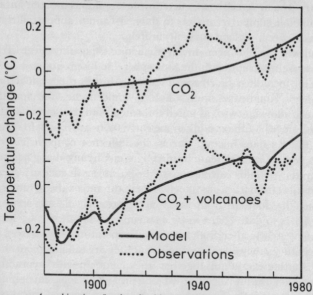

Figure 10.2. A combination of carbon dioxide greenhouse effect and volcanic dust veiling explains at least some of the recent pattern of temperature changes. But clearly other factors are also at work. (Source: *As Figure 10.1.*)

colleagues say that the global warming for a fast-growth world ('business as usual') would be close to 4° C by the end of the twenty-first century; slow growth reduces the forecast warming to about $2\frac{1}{2}$° C, and zero growth (or slow growth in energy use with coal phased out early in the twenty-first century) causes a rise of only 1° C over the next 120 years. 'The predicted CO_2 warming rises out of the ... noise level in the 1980s ... [this] does not depend on the scenario for atmospheric CO_2 growth, because the amounts of CO_2 do not differ substantially until after year 2000.' Volcanic eruptions on the scale of Krakatoa or Agung might delay the warming, but only by a few years, and if the forecast stands up, modellers ought to be able to confirm the reality of the greenhouse effect with 85 per cent confidence (in statistical terms) in ten years' time, and 98 per cent confidence by the end of this century.

If the computer models are correct, and if forecasts like that of Wallace Broecker are taken at face value, by the year 2000 the world could be in a climatic state it has not experienced for the past 1,000 years. That does not represent a disaster scenario, since our ancestors of a millennium ago managed quite happily. But is modern agriculture as flexible as the peasant cultivation of 1,000 years ago? The tendency of developments in scientific agriculture during this century – especially in the second half – has been to produce crop strains and farming techniques which produce very high yields under optimum conditions. But the 'optimum' conditions those crops and techniques are tailored for are usually the 'normal' conditions of the years when the crops and techniques were being developed – the 1950s and 1960s. The greatest concern about the prospect of an imminent climatic change, in any direction, comes from those people who see modern agriculture as tailored precisely to the normal weather of the past two or three decades. We may have gained productivity at the expense of flexibility, the argument runs, so that while the good years are very good, any deviation from perfect weather conditions causes a dramatic reduction in yields. And, all the while, world population is increasing. The pattern of regional variations in the weather that forecasters expect to be associated with any global warming due to the greenhouse effect only highlights the importance of these problems, and suggests that the consequences of the greenhouse effect may hit us before any global warming does become detectable against the background noise – may, indeed, already be upon us.

The regional implications

Global temperature averages tell only part of the story. The model used by Manabe and Wetherald, for example, predicts, for a doubling of the atmospheric carbon dioxide concentration, very little warming in the tropics, a $10°\,C$ increase in polar temperatures, a warming by $2–3°\,C$ in middle latitudes, a 2 per cent increase in relative humidity and a 7 per cent increase in average precipitation. To investigate details of the regional changes, though, we have to leave the computer models to one side and look again at the way the weather changes from decade to decade in the real world.

The key feature of the GCMs is that they predict a rapid increase in global temperatures of a degree or more within a human lifetime. Although Idso disputes this, the key feature of his approach is that it tells us there is an absolute limit to the increase, which might be reached in a century or two. They could both be right – a rapid increase of a degree or more could be followed by a slower rise up towards Idso's limit (if fossil fuel continues to be burnt) as the last panes in the global greenhouse are filled in. Even the best computer models cannot do more than provide an outline sketch of how different regions of the globe will be affected by the changes associated with any such increase in mean temperatures, but there have been warm and cold years, worldwide, from time to time during the past fifty years or so, the period for which we have good records of temperature, rainfall and so on from many sites around the globe. Several climatologists have been attempting, in the early 1980s, to use these records to build up a picture of what a 'warm Earth' would be like. Their idea is simply that by combining the records of several warm years they might produce a picture of a 'typical' warm pattern, while the records from colder years might provide a similar 'cool Earth' composite. Comparing the two ought to give a good guide to the *direction* of changes associated with the greenhouse effect, even if the size of those changes cannot be predicted in detail. And the trick seems to work.

Two groups in particular have pioneered this kind of study. Jill Williams at NCAR and a team headed by Tom Wigley at the University of East Anglia in the UK have made broadly similar approaches to the problem and come up with broadly similar answers; since these answers also agree with the limited information available from the computer models, but extend it further, it looks very much as if they are both on the right track. Both these studies, however, show up some important differences between small changes in global temperature in the twentieth century and the large changes in temperature that have happened in the recent history of the Earth. The nature of those differences is not reassuring for any residents of Europe or North America who have a gut feeling that a slightly warmer world might be a nicer place to live in.

First, the broad perspective. Will Kellogg, also of NCAR, has gathered together the evidence from tree-rings, pollen remains, isotope studies and other geological records, which shows that there have been four epochs during the past $2\frac{1}{2}$ million years which might be regarded as representing four progressively warmer states of the Earth. These are the Medieval Warm Period, also known as the Little Optimum, from about AD 800 to 1200; the Hypsithermal, or Altithermal, between about 4,000 and 8,000 years ago; the previous interglacial, centred on a date about 120,000 years ago, which lasted for about 10,000 years; and the last time the Arctic was entirely free of ice, a warm period extending from about $2\frac{1}{2}$ million years ago back for at least hundreds of thousands of years. Kellogg chose the Hypsithermal period for a detailed reconstruction – as detailed as possible from the evidence of animal remains, sediments and so on – which might indicate the sort of changes we could be in for in a greenhouse world. His reconstruction (Figure 10.3) shows that between 4,000 and 8,000 years ago North and East Africa was more favourable for agriculture than it is today, Europe was wetter, and Scandinavia and the present grain-growing regions of North America were drier.

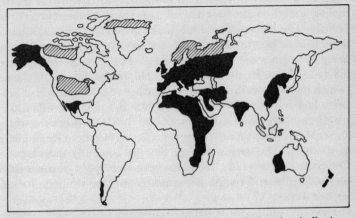

Figure 10.3. Will Kellogg's reconstruction of rainfall distribution when the Earth was warmer than today, during the Hypsithermal between 4,000 and 8,000 years ago. Some parts of the world (black) *were wetter than today, others* (shaded) *were drier. Information is lacking for blank areas. (Based on Figure 13.6 of W. W. Kellogg, 'Global influences of mankind on the climate', in* Climatic Change, *ed. J. Gribbin, Cambridge University Press, London and New York, 1978.)*

If the anticipated global warming produced a similar pattern of rainfall, North America would produce less grain, but parts of North and East Africa, the Middle East, Mexico and Western Australia, all of which now suffer frequent drought, could become grain exporters. Such a scenario has interesting political implications, since today the United States exerts considerable influence on the world not just through its military strength but also as a provider of grain both to the Third World and to the USSR. The implications of the scenarios derived from twentieth-century records are, however, even more 'interesting'.

Taking the University of East Anglia study as representative of the results obtained by all the teams using this kind of approach, we can look in detail at the scenarios. The team looked first at the temperature records from around the Arctic, in a zone from 65° N to 80° N, and took a composite of the five warmest years in the interval 1925–74 and of the five coldest years in the same 50-year period, using them to develop a picture of the differences between a cold Earth and a warm Earth. The five warmest years at high latitudes were 1937, 1938, 1943, 1944 and 1953; the five coldest were 1964, 1965, 1966, 1968 and 1972. This brings out very clearly the global cooling after the 1940s, but that, of course, was not the object of the exercise. The University of East Anglia team looked first at a high-latitude zone because the GCMs tell us that higher latitudes are likely to warm more than the average. The average temperature difference between the warm and cold composites in this zone (which includes northern Norway, northern Sweden, much of Finland, Iceland, northern Canada, Alaska and northern Siberia) is 1·6° C, compared with a difference of only 0·6° C between the warm and cold composites for the same years for the Northern Hemisphere as a whole. This seems to bear out the computer calculations. Comparing winters only, the warm years were 1·8° C warmer than the cold years, and comparing summers only the temperature difference was 0·7° C.

Figure 10.4 shows the pattern of temperature changes between the warm and cold five-year composites in more detail. Maximum warming occurs in the continental interiors at high latitudes. A region from Finland across Russia and Siberia to 90° East warms by up to 3° C, much of North America shows increases of 1–2° C, but some regions – Japan, India, Turkey and Spain among them – actually show a decrease in temperature. These changes are explained meteorologically by shifts in the atmospheric circulation pattern which bring an intensified westerly air flow and deeper depressions between 50° and 70° N, and a westward displacement of the Siberian High in winter.

The warm years also show a 1–2 per cent increase in overall rainfall compared with the cold years, but once again this modest overall average change conceals bigger regional effects (Figure 10.5). There is a *decrease* in rainfall over much of the USA, Europe,

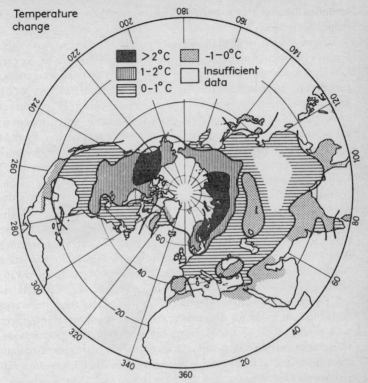

Figure 10.4. The University of East Anglia team's reconstruction of the patterns of temperature change between warm and cold years in the period from 1925 to 1974. Any global warming in the near future seems likely to produce a bigger effect at high latitudes than closer to the Equator. (Data supplied by Tom Wigley.)

Russia, and Japan, combined with a bigger-than-average increase over India and the Middle East. Seasonal variations are as important as geographical ones, with a more vigorous monsoon circulation in warmer years; the seasonal temperature patterns show, for example, that while North America warms in all seasons, the cooling around Turkey is related to a decline in spring temperatures, while the cooling in India is greatest in summer and autumn.

Overall, the most important differences between this study (and Jill Williams's analysis) and the Hypsithermal scenario from Will Kellogg are that Europe and Russia are drier, and Canada wetter, than Kellogg suggests. Crudely speaking, this is a bad thing for the

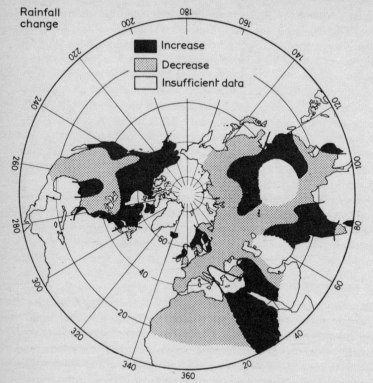

Rainfall change

■ Increase
▨ Decrease
□ Insufficient data

Figure 10.5. The pattern of rainfall changes associated with the temperature changes indicated in Figure 10.4, for a change from colder to warmer Earth. The prospect of drought in North America, across Europe and into the USSR is one of the most disturbing prospects if the greenhouse effect takes a grip on our planet. (Data supplied by Tom Wigley.)

present pattern of world food – especially grain – production, although we shall look in detail at the agricultural implications in Chapter 11. Warmer and drier conditions would not help the grain-growing regions of either North America, the producers of the present world surplus, or the USSR, which is already unable to produce enough grain for its present needs. No wonder the carbon dioxide greenhouse effect has become such a political hot potato.

But the University of East Anglia study also brings out other points. Even in a warmer world, there will still be year-to-year variations in temperature and rainfall, with some years being hotter than others and some cooler, some wetter and some drier. Wigley and his colleagues emphasize that in any case, like all such studies, theirs provides only a scenario of what the world might be like if it warmed slightly, not a prediction of what it will inevitably be like – the reasons which made some years between 1925 and 1974 warmer than others were not directly related to a build-up of carbon dioxide in the atmosphere. Two other researchers at the University of East Anglia's Climatic Research Unit, Chris Sear and J. M. Lough, have examined all the present crop of warm world scenarios, and they conclude that while they are valuable tools as long as they agree with each other and with other forecasts – from the computer, for example – a great deal more work of this kind needs to be done. After all, for parts of Europe the instrumental record of weather goes back to the 1780s, and there have been several spells of warm and cold weather since then. The potential for this approach to the problem is great, but as yet it has barely been tapped.

Why people are worried

As we shall see, a combination of increased temperature, increased rainfall and a greater concentration of carbon dioxide in the atmosphere might actually be beneficial for farming in many parts of the world, especially the hungry regions of the Third World where rice is the staple crop. But the present-day world trade in food is still almost entirely dependent on the surplus produced in the grain-growing regions of North America. The US–Canadian surplus is the world's insurance against famine, and if conditions change millions could starve during the period of adjustment, even if in the end more food could be grown. Such dangers are highlighted by the evidence that even today the world could provide enough food for twice the present population, yet millions still live on the edge of starvation. People who talk about what may be possible in a warmer world must remember that the practical achievements of the world food production and distribution system today fall far short of what is, in theory, possible, as has been spelled out by many groups around the world, including a team at the UK's Science Policy Research Unit, working with Christopher Freeman and Marie Jahoda.

If agriculture were already being run effectively, to provide a healthy diet for the maximum number of people, then none of the climatic changes likely to occur in the next fifty years, for natural reasons or through the activities of mankind, would pose any problem. But in the real world, a reduction of 10 per cent in North American harvests would cut off, at a stroke, all of the surplus available to hungry people in other parts of the world. This is the most immediate, and realistic, hazard associated with the greenhouse effect.

A less immediate, and less realistic, but in some ways more widely publicized hazard is the threat of inundation of coastal regions by rising sea-level as the polar ice-caps melt. This is a dramatic possibility, but one that is, in reality, rather remote. First, the certainties. Quite apart from the effects on ice, an increase in temperature would cause the ocean waters themselves to expand, raising sea-level by a metre for a warming of 5° C – which at worst would not happen for centuries, and at best may be impossible. In any case, it would take a very long time to warm the oceans, even if the temperature of the troposphere did increase by 5° C. Next, when floating ice melts there is no change in sea-level, because the floating ice displaces as much water as it releases when it melts. So whatever happens to the ice cover of the Arctic Ocean, we don't have to worry. The hazard, such as it is, comes from the great mass of ice grounded on the Antarctic continent.

Normally ice melts slowly, and it would take many centuries for the Antarctic ice-cap (or the smaller ice-cap on Greenland) to respond to a global warming – centuries in which the greenhouse effect might be brought under control, cooling the Earth again before the sea-level rose significantly. But the West Antarctic ice-sheet extends far out over the sea, much of it grounded on the floor of what would otherwise be the sea-bed. Under those conditions, some theorists believe, a relatively small change in temperatures and rise in sea-level could allow ocean water to push under the ice-sheet, breaking it up by 'calving'. As ice broke away from the edges, the weight of the ice-sheet would be reduced, causing it to spring higher away from the sea-bed, letting still more water underneath. The whole West Antarctic ice-sheet might then break up in an 'ice surge'. There is good evidence that 120,000 years ago, in the interglacial before the present one, some sea-levels were 6 metres (20 feet) higher than today, and this is probably because the West Antarctic ice-sheet had broken up then. A few glaciologists – notable among them J. H. Mercer of the Institute of Polar Studies, Ohio State University – believe that a surge of the West Antarctic ice-sheet could occur within the next fifty years. Most of his colleagues disagree and point to the Hypsithermal as a period of global warmth, compared with the present day, when the West Antarctic ice-sheet definitely did not break up.

The 'threat' of such a break-up, however, makes for dramatic headlines, with the implication that Florida, Texas and the Gulf of Mexico would be inundated, with millions of people affected at a cost of well over a hundred billion dollars. With many nuclear reactors sited along coasts, as well as major cities such as London, Leningrad, New York, Washington DC, Amsterdam, Venice, Calcutta and Singapore, a West Antarctic ice surge seems to have all the ingredients of a great global disaster. The truth – even if it happened – would be a little more mundane.

Stephen Schneider and Robert Chen have carried out a study of the implications for the USA. A rise in sea-level of 5–8 metres occurring over a year or two would affect 5·7 per cent of the population of the continental United States directly, and almost wipe Florida from the map. But its effect on GNP, Schneider and Chen estimate, would be only slightly greater than that of the oil embargo of 1974. Such a rise in sea-level over the span of a decade – still dramatically rapid by glacial standards – would, they say, be 'less of a shock'. The implication is that if the USA coped, albeit with some difficulty, with the 1974 oil crisis, then the world would cope even with a surge of the West Antarctic ice-sheet. Against this background, doomsayers who argue that there is an urgent need to take steps to control carbon dioxide emission before the world suffers irreparable damage have to make a case that the cure would not be worse than the disease. It might, in fact, be less of a strain to the world economy and socio-political systems to let the climatic change happen and learn to live with it, than to try to learn to live without using more fossil fuel.

Burn less?

One of the problems about trying to plan ways in which to deal with the build-up of carbon dioxide – the 'technofix' solution – is that there is so much uncertainty about the likely growth of global energy demand, and the mix of fuels that will be used, over the next 50 to 100 years. Many pre-1975 projections now look downright silly, but even today the energy experts can offer no clear guidance about the future. This is a book about weather, not energy forecasting, and more expert authors than I have filled much larger books than this with their attempts to provide a reliable basis for energy planning. But from this mass of – sometimes contradictory – information it is possible to pluck two scenarios to give us a feel for the range of possibilities now facing the world. I have chosen these representative 'best' and 'worst' scenarios from a series of projections made in 1978 by a team at the International Institute for Applied Systems Analysis (IIASA) in Austria; these are not intended to represent the most accurate or reliable guides around, but to indicate the sorts of forecasts the energy experts play with.

In the mid 1970s the total world consumption of energy was about 7·5 TeraWatts (1 TW = 1 million million watts). Of this, the developed countries used about 5·3 kiloWatts (kW) per head, and the developing countries about 0·45 kW per head. Forecasts of future energy demand depend both on assumptions about the amount of energy used by each person in the world and on assumptions about the number of people in the world. Several studies have now shown that with only modest efforts made to use energy more efficiently, the rich North could maintain a rising standard of living with no increase in the amount of energy used per person, well into the twenty-first century. One such study, concentrating on the United Kingdom but relevant to all developed countries, was published by the International Institute for Environment and Development in 1979. A team headed by Gerald Leach showed how, even if GDP trebled by 2025, simple, proven technology could ensure no increase in total energy demand for the UK.

These projections are not based on any belt-tightening or doing without. They assume what now seem rather rosy 'business as usual' futures, with increased ownership of freezers, dishwashers, colour TV, cars and so on. The energy savings come from more efficient use of fuel, for example in private cars (where there have already been significant improvements), insulation of buildings, and so on. Such a forecast has dramatic implications for planning, not the least being a saving of about £30,000 million over a thirty-year period, compared with present UK Department of Energy forecasts, by not building so much generating plant. This saving would contribute a major part of the cost of energy efficiency measures – and nuclear power scarcely enters into the picture at all.

In the real world, things are different. At the time of writing, the

UK has a government apparently enthusiastically pushing ahead with the nuclear programme, while the policies of that government have produced such a sharp slump in productivity and industrial activity that the IIED forecasts, described as 'business as usual' only two years ago, now look like the wild-eyed extreme of high growth. Experts today are now projecting an immediate future of almost no energy growth, while by the end of the 1970s actual energy consumption in the UK was already below the consumption at the beginning of the decade, in annual terms. Similar sweeping changes are occurring in Europe, and the United States is now beginning to follow suit. Against this background, and with an awareness of the need to use energy efficiently now instilled in every planner in the developed world, there is no reason to expect the per head consumption of energy in the rich North to rise in the immediate future. Assuming the poor South will aspire to the same level of consumption, the most reasonable 'guesstimate' of the demand for energy in a fully developed twenty-first-century world is 5 kW per person per year.

The IIASA projections use just such a guideline. To cover a range of population growth figures, and allow for some uncertainty in the estimates, the IIASA study picks out a 'low' energy requirement of 30 TW, and a 'high' demand of 50 TW, for the year 2025. These are, respectively, four times and almost seven times the global demand for energy each year in the mid 1970s. After about 2025, energy requirements should rapidly level off as population starts to stabilize – lower birth rates are a feature of all developed countries, and this energy demand assumes a fully developed world. The alternative, unpleasant, possibilities do not lead to any energy problems. Large-scale nuclear war, or mass starvation in the poor South, would both 'solve' the energy problem at a stroke.

Each of these energy scenarios could be met by a variety of fuels in different combinations, but, again in the search for simplicity, I shall pick out the two most extreme cases in terms of carbon dioxide build-up. One is heavily dependent on nuclear power, the other heavily dependent on fossil fuel. So, in terms of a carbon dioxide build-up, the 'worst case' is the 50 TW energy scenario, with heavy dependence on fossil fuel. In this scenario, carbon dioxide concentration climbs throughout the twenty-first century, eventually reaching at least eight times the pre-industrial concentration. Global temperatures increase dramatically after about 2020, reaching 5°C above present levels in 2060 and 8°C above by 2100. (All this assumes, of course, first that half the carbon dioxide released stays in the air, and second that the GCMs provide a good guide to the temperature changes that result.)

At the other extreme, a 30 TW per year world, with heavy reliance on nuclear power, would be unlikely to produce any noticeable greenhouse effect, with the carbon dioxide concentration never reaching twice its pre-industrial level (see Figures 10.6 and 10.7). Neither scenario is realistic, of course, and all the assumptions are questionable. The truth, in terms of carbon dioxide build-up, probably lies somewhere between these extremes. Faced with such uncertainties, it is small wonder that the reaction of many planners is to ignore the problem. And they may be right to do so – or, at least, to avoid the trap of devoting money and research effort into a major attempt at a technofix for the carbon dioxide build-up.

The cost of control

There are, however, policy implications in the greenhouse effect that do provide immediate food for thought. The most obvious, perhaps, is the case of the recent enthusiasm amongst US policy-makers for 'synfuels', coal converted to oil and gas. For each unit of energy produced, synfuels release 2·3 times as much carbon dioxide as natural gas, 1·6 times as much as oil, and 1·4 times as much as coal. That is the sort of number which ought to enter into the policy-makers' calculations, if only to tilt the balance where a marginal decision has to be made on whether to go ahead with synfuels or some other, almost equally attractive energy source. The real technical fix, endorsed by a few technological optimists, involves rather more than this, however.

If carbon dioxide in the atmosphere is perceived as a bad thing, then there are two ways to tackle the problem. One is to produce less – opt for nuclear power rather than synfuels, perhaps, or improve efficiency and concentrate on renewable sources (but try telling that to the Third World). The other alternative is to find ways of extracting carbon dioxide from the air, or from power station smokestacks, and storing it indefinitely. This is the technofix.

One of the ironies of the carbon dioxide build-up is that the developed countries which started the ball rolling are not going to be the main offenders over the next few decades, but are in the best

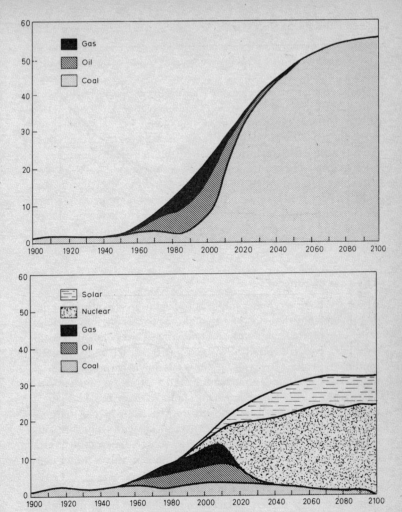

Figure 10.6. Two energy scenarios contrast the nuclear and fossil fuel 'options'. The upper diagram is a 'worst case' scenario for the build-up of carbon dioxide, with world energy demand growing to 50 TW, seven times the 1980 level, and most of it coming from coal. The lower diagram is the nuclear option, with solar power playing a secondary role and fossil fuel use declining in the twenty-first century. Peak levels in this scenario only reach 30 TW. Clearly, many other scenarios can be imagined; these two have been chosen simply to highlight the extremes of choice open to mankind. (Data from IIASA studies, supplied by Jill Williams.)

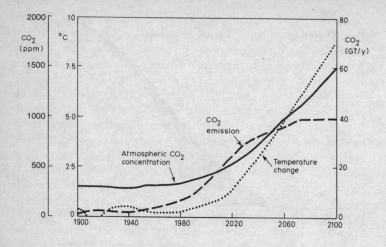

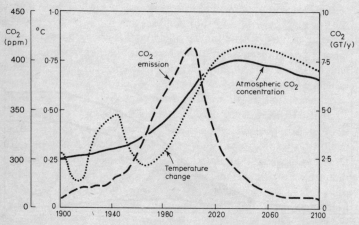

Figure 10.7. Using the standard computer models of the carbon dioxide greenhouse effect, it is possible to calculate the global warming associated with each of the energy scenarios of Figure 10.6. In the 50 TW fossil fuel scenario (top) *carbon dioxide emission and temperature keep on rising throughout the twenty-first century. In the nuclear/solar 30 TW scenario, carbon dioxide emissions drop rapidly after the year 2000 and temperatures never rise more than half a degree above the 1980 level. Note the difference of scale between the two figures.* (Source: *Jill Williams, IIASA.*)

position to develop a technofix. But removing carbon dioxide from the air on a large enough scale to reduce the overall atmospheric concentration does not seem to be practicable. The chief problem is that, even with the build-up due to man's activities, the concentration is still so low – a few hundred parts per million, 0·03 per cent. In the sixth report of the US DoE Carbon Dioxide Effects Research and Assessment Program, published in 1980, Anthony Albanese and Meyer Steinberg showed that significant quantities of carbon dioxide could only be removed from the air by using powerful fans to suck in enormous quantities of air for treatment. The power to drive these fans has to come from somewhere, and if the source was electricity from a fossil fuel power station as much carbon dioxide would be released from fuel burnt to supply that energy as would be extracted from the air by the processing plant. Similar arguments apply to removing carbon dioxide from the air by freezing – it can be done, but the energy costs are far too high.

It is easier to remove significant amounts of carbon dioxide from the flue gases of large power stations, but that still leaves the problem of what to do with it. One plan envisages the collected carbon dioxide being injected into the deep ocean, perhaps through a pipe which extends into the layer of water flowing out of the Mediterranean Sea through the Straits of Gibraltar. This bottom current dives deep into the Atlantic Ocean, joining a ponderous deep ocean circulation system which would take thousands of years to bring it to the surface. But even though the concentration of carbon dioxide in flue gas is about 500 times greater than in the atmosphere, the DoE study points out that it is still difficult to encourage large quantities to dissolve in sea-water. The system which the study pinpoints as using least energy involves removing carbon dioxide from flue gases using an absorbent called monoethanolamine (MEA). Carbon dioxide is driven off from the MEA by heating and stored; the MEA is recycled. And concentrated carbon dioxide obtained in this way could easily be transported and disposed of, either as a liquid or a solid. Solid carbon dioxide, dry ice under modest refrigeration, could be taken by barge out into the deep ocean and tipped over the side. Because it is denser than water, the dry ice would sink, eventually melting and bubbling off gas to dissolve in the deep sea, out of contact with the atmosphere for millennia. In liquid or gas form, the carbon dioxide could be disposed of through a long pipeline from a power station out to the deep ocean, some 150 km (100 miles) offshore in the case of the USA. The important factor in assessing the practicality of such schemes is not so much the cash cost but the energy cost – and this is where they fall down.

According to the DoE study, to remove and dispose of half of the carbon dioxide in flue gases by MEA scrubbing would reduce the efficiency of the power station by about 50 per cent, and therefore increase the cost of electricity correspondingly. For 90-per-cent carbon dioxide removal, the power produced would become at least three times as expensive, perhaps more.

There is an opposing camp in this debate. Cesaro Marchetti at IIASA has calculated that scrubbing flue gases to remove carbon dioxide, using similar techniques to those used to remove sulphur dioxide today, could be done at a cost of only a few dollars per ton of carbon collected, up to the point where 50 per cent of the carbon dioxide was being trapped. C. Mustacchi and colleagues of the Analysis and Development of Energy Systems group in Rome have made the even more dramatic claim that 'no major difficulty and only moderate cost will be involved' in removing carbon dioxide from flue gases. The overall additional cost of electricity generated would, they calculate, be no more than 20 per cent.

Even if the optimists turned out to be right, their comforting calculations lose sight of the fact that in the real world energy growth in the immediate future will be concentrated in the Third World, not in the rich North where the technological expertise and, perhaps, willpower to tackle a technofix for carbon dioxide exists. An additional cost of 20 per cent for energy represents an enormous burden for a developing country, and it is hard to see any genuine prospect of carbon dioxide being controlled by extracting the gas from power station effluent. Some of the other ideas which surface from time to time are even more extravagant and unrealistic. Deliberately lacing the atmosphere with dust – an artificial volcano – to block out solar heat, or scattering reflective plastic discs on the ocean surface to reflect away solar energy and cool the Earth, are just two which have been put forward, apparently seriously, at scientific conferences in recent years. But the voice of reason still prevails in most scientific debate on the issue, and gives some cause for hope.

Reasons for hope

The cost of controlling carbon dioxide emissions on a global scale is too high to pay. Without a cripplingly expensive technofix, it implies stunting the growth of the developing nations and leaving the poor of the world to get poorer while the rich continue to do very nicely. But that does not mean there are no steps to be taken to minimize, relatively painlessly, the cost of any carbon dioxide build-up. One approach is actuarial – or, as Stephen Schneider puts it, 'hedging our bets'. 'Rather than gambling our energy future with an ante of 88 billion in synfuels,' he says, 'why not hedge our bets with, say, 8·8 billion invested in each of ten different options?' Roger Revelle and Donald Shapero summed up the pros and cons of the technical fix options in a 1978 paper, and concluded that 'it may well be the case that increasing reliance on renewable resources, with the concomitant reduction in the carbon dioxide burden in the atmosphere, will emerge as a more practical alternative'. It is no bad thing if concern about the carbon dioxide build-up makes us more cautious in our use of fossil fuel, both at national and global levels, and there are certainly alternative energy sources not just on the horizon but now looming large in the foreground.

Alcohol made from corn or maize is now a practical fuel for cars; a breakthrough by Swiss researchers in 1980 encouraged Nobel Laureate Sir George Porter to make the prediction that artificial photosynthesis, splitting water into hydrogen (for fuel) and oxygen by photochemical means, could become a practical reality within a decade; and in Holland researchers at the Institute for Atomic and Molecular Physics reported in 1981 the development of new techniques which drastically reduce the cost of manufacturing solar cells to generate electricity.

It is entirely on the cards that before the end of the present century the carbon dioxide problem will have disappeared, not because of any concerted attack on the problem by science, technology and policy-makers, but because alternative energy sources have been developed for much more pressing economic reasons. Perhaps the greatest threat to this prospect lies not so much in the demands of industry and households for energy, but in the demands of people for food – and not just the essential food necessary to keep body and soul together. One of the great unknowns in the whole carbon dioxide debate is how much the destruction of tropical forests is contributing to the build-up of carbon dioxide in the atmosphere. There are many reasons for preserving what remains of this unique biological resource; in the present context, what matters is that tropical rain forests alone hold 42 per cent of all the carbon that is now held by terrestrial vegetation, and all the forests of the world put together account for about 90 per cent of vegetable carbon, even though dense forests cover only about 30 per cent of the world's land surface. At the present rate of destruction, there will be no tropical rain forest left by the end of the present century, and that complete removal of one piece of the environmental jigsaw puzzle might have far more dramatic repercussions than any release of carbon dioxide from burning fossil fuels. But the forest is not being destroyed to feed the hungry mouths of the Third World. In what Norman Myers calls 'the hamburger connection', nearly 40 per cent of the forest cover in Central America has been turned into pasture to feed the cattle that in turn end up in the products of North America's fast food industry. Economizing on food – or at least avoiding waste – might be an even more practical contribution to solving the carbon dioxide problem than economizing on energy, at least for those of us living in the rich North. And agriculture may provide the solution to the carbon dioxide problem in other ways.

After all the fuss, the best climatic forecast we can make for the immediate future – to the end of the present century – remains to expect what has happened before, with year-to-year variations in the range that has been covered in the past 200 years. If there is a greenhouse-effect contribution to rising temperatures and regional shifts in both temperature and rainfall distribution, then those effects will be felt first by the farmers. So any efforts to minimize the impact of a carbon dioxide build-up in the atmosphere should be devoted to improving the efficiency, and ruggedness, of agriculture. This approach has the benefit that if the doomsayers are wrong and the climate does not change noticeably, the improvements in agriculture will still be of immense value in a world where population is still increasing rapidly. The worst thing that could happen if we opted for a technofix would be the investment of billions of dollars in technology that proved unnecessary; the worst thing that can happen if we boost agriculture is that fewer people starve. In a world where the climate varies on all time-scales from decades upwards, the single most important investment today must be in agriculture. If the carbon dioxide build-up increases our concern about climatic change, that is simply one more reason to invest in improved food production, especially in the Third World. And there is great potential to improve agriculture to the point where the kind of climatic change likely to happen in the next hundred years poses no threat at all.

11

A Thought for Food

To the ordinary person trying to make head or tail of the debate about future weather trends, the prognostications of the experts may seem to differ so wildly that there is a natural impression that none of them really know what they are talking about at all. The problem, though, is really to do with the specialization of science today, the way experts are compartmentalized so that few of them have any grasp of the broad picture. This problem is highlighted by the way in which the initial reaction of all climatologists to the possibility of a build-up of atmospheric carbon dioxide was to look at the effects on agriculture in terms of droughts, floods and so on – the physical effects which they, as physical scientists, were familiar with. Only when the biologists and agricultural scientists heard the news and responded to it did the community at large receive the, perhaps crucial, reminder that one of the other effects of increasing carbon dioxide concentration in the atmosphere could be to increase the efficiency of photosynthesis in plants, improving crop yields and even improving the efficiency with which crop plants make use of available water. It isn't that either group of specialists is 'wrong' in their assessment of the problem, just that they have different perspectives on a complex issue.

Even though the issues remain complex, however, enough experts with their different points of view have now contributed to the debate for an outsider with no particular axe to grind to be able to pick a way through the arguments and arrive at some kind of overall picture. This may not yet be the final picture – a great deal more work needs to be done, first to establish the size of any greenhouse effect and second to establish how that change will affect world agriculture. But while the details of the picture may change, the outlines are likely to remain much the same. And the outlines can be sketched in by looking at the different opinions and the reasons why such a range of opinion exists. Since the pessimists and doom-mongers have been responsible for most of the action – both scientific and in the public debate – on the greenhouse effect in the past decade, perhaps they have earned the right to pride of place in this survey.

The pessimists

How bad could the greenhouse effect be for world agriculture in the immediate future? Any such assessment depends on how well we can forecast not just the likely rise in temperature, but also the patterns of rainfall in a warmer world. So far, the only way to do this, as we have seen, is by comparing rainfall patterns from different times in history – and pre-history – when the average temperature of the world was different from today. The critics of this approach argue that we don't know *why* the Earth was warmer at certain epochs in the past, and that a greenhouse-effect warming might produce a different rainfall pattern than, say, a slight increase in the heat output from the Sun. But some of the force of that argument has been removed by the studies of air bubbles trapped in polar ice, which show how the carbon dioxide concentration of the air has changed over the past 30,000 years.

Robert Delmas and his colleagues reported in 1980 that at the height of the most recent Ice Age there was only about two thirds as much carbon dioxide in the air as there is today, while other studies show that about 5,000 years ago there was considerably more, perhaps as much as 500 ppm for a short period. S. L. Thompson of the University of Washington, and Stephen Schneider, whom we have met before, summarized the immediate implications of these discoveries in *Nature* in March 1981. The changes in carbon dioxide concentration are not enough to explain the shift from Ice Age to interglacial and back again, but still there are enormous implications for the study of past climates. Most probably, as the Earth cools for some other reason – such as the Milankovitch process – the cooler sea is able to take up more carbon dioxide, allowing the Earth to cool still more as the atmospheric greenhouse effect is reduced. Such a feedback could explain half of the change in surface temperature of the globe from Ice Age to interglacial. And if there was more carbon dioxide in the atmosphere 5,000 years ago – for whatever reason – then using Will Kellogg's reconstruction of the Hypsithermal rainfall patterns may not be such a bad guide to a greenhouse world after all.

With that in mind, the pessimists argue that the world agricultural system is today so finely tuned to present climatic patterns that any change in what we think of as 'normal' weather must be bad, while the scope for disaster increases the more 'artificial' agriculture becomes. W. Bach of the Centre for Applied Climatology, based in Munster, West Germany, is among the experts who have pointed out the gloomy implications.

As Figure 11.1 shows, the temperature in the US corn (maize) belt today is exactly right to produce maximum yields of the crop. Any change in summer temperatures would reduce the yield, and any decline in July rainfall would have a similar effect. A reduction of one inch (25 mm) in July rainfall corresponds to a 7-per-cent reduction in yield, and summers that are both drier and hotter are bad news for farmers. If summers are both wetter and cooler, the benefits of increased moisture outweigh the losses due to increased temperature; but all the climate reconstructions point to warmer and drier conditions in this critical region of the world as the greenhouse effect develops, with each $1°$ C rise in temperature producing a loss of 11 per cent of the yield, in addition to the losses caused by drier conditions.

Wheat, another major crop, needs an unusually long growing season compared with other grains, is very sensitive to extremes of both heat and cold, and requires relatively high minimum temperatures. Any increase in temperature is likely to reduce yields, although the exact effects vary even from state to state within the continental US. In other parts of the world, notably Kazakhstan, Bach predicts that if temperature increased by $1°$ C and precipitation declined by 10 per cent, wheat yields would be reduced by as much as 20 per cent. Of course, the grains from the temperate regions of the world are not the only crops. One third of mankind depends on rice for more than half of its food, and as we shall see the implications of a greenhouse-induced climatic change may be very different for crops such as rice grown in warmer, wetter regions of the globe. But the pessimists argue that the effect of any reduction in US and Canadian grain production would be devastating for the world food market. Prices would rise, and the surplus available for food aid would disappear, leaving millions to starve before they had any opportunity

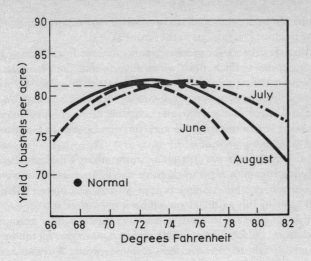

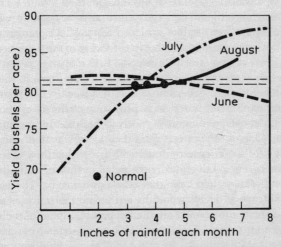

Figure 11.1. Corn yields in the US corn belt are critically tuned to the present 'normal' summer temperatures (top). *Yields are less critically sensitive to rainfall, but a combination of warmer and drier conditions – exactly the pattern the greenhouse effect is likely to produce – could have a severe impact on the crop.* (Source: *R. H. Biggs and J. F. Bartholic, 'Agronomic effects of climatic change',* in Proceedings of the Second Conference on CIAP, *ed. A. J. Broderick, US Department of Transportation, Washington D.C., 1973.*)

to change their way of life to take advantage of any improvement in agricultural conditions in the Third World.

Almost every week our newspapers carry reports of famine somewhere in the world, reports which suggest that the pessimists have reason for their concern. To take just one example, on Wednesday 8 April 1981 the *Guardian* carried a story reporting Vietnam's appeal to the United Nations for 384,000 tons of emergency food aid, following drought, typhoons, and 'the worst floods in the country's food-producing provinces for 30 years'. There is nothing special about this example, except for the irony that, whether or not the appeal is routed through the UN, the food sent as aid most probably comes from the USA, until recently Vietnam's bitterest enemy. In that part of the world alone we have seen similar requests from China, massive aid to Kampuchea, and many smaller examples of food from the temperate north going to hungry mouths in the allegedly fertile tropical and semi-tropical regions. A decade ago, the great hope was that science would solve all these problems through the 'Green Revolution', the development of new strains of high-yield crops tailored to the needs of agriculture in the developing countries. But perhaps the most potent shot in the locker of the pessimists today is the failure of that Green Revolution to live up to expectations.

The first seeds of the Green Revolution were sown in the 1940s, when the Mexican government and the Rockefeller Foundation together established a research programme aimed at increasing Mexican wheat production as much as possible and as quickly as possible. One of the key problems they faced was that when traditional wheat varieties were dosed with large quantities of fertilizer, the plants grew so tall and carried such a burden of grain that they became top heavy and bent under their own weight. The answer to this problem was developed in the form of new wheat varieties with short, stiff stems that could take 120 pounds of nitrogen fertilizer per acre without collapsing, compared with a 40-pound-per-acre limit for the older varieties. By the mid 1950s Mexican wheat yields had doubled, and they doubled again by the mid 1960s. At the start of the research programme Mexico imported half of the wheat its people needed; by the late 1960s it was a grain exporter. And by the 1970s, with explosive population growth, Mexico was importing food again.

This is the classic example of increasing population outstripping increasing agricultural productivity, and one the pessimists often turn to. Yet, for the world as a whole, over the same period from the 1940s to the end of the 1970s, food production worldwide rose at an average 2·8 per cent a year, while population grew at an average 2 per cent a year. There are limits to growth, but the past forty years have not seen us nudging up against them. It is other problems that have, so far, stifled the original optimism about the Green Revolution.

High-yield varieties of wheat and the development of so-called 'miracle rice' seemed to have solved the immediate world food problem when countries such as India and Pakistan, the Philippines, Mexico, Indonesia and Sri Lanka achieved dramatic increases in food production with their aid. The new varieties certainly do well under ideal conditions. But it now seems they may be less productive than the old varieties unless they receive just the right inputs of water, fertilizer and sunshine. At the same time, dwarf grain varieties with heavy heads are almost a gift to pests, such as rats, who get a free meal thanks to mankind's ingenuity. The Green Revolution certainly did increase yields, and it certainly saved many people from famine. But at best it bought a breathing space, a little time for the people of the world to tackle the real problems, chiefly increasing population. The 'failure' of the Green Revolution is really the failure of the world to use that time – scarcely more than a decade – even to begin to bring population growth under control. By inducing a sense of complacency, a feeling that science had solved the food problem, the Green Revolution may even have been a bad thing, the most extreme doom-mongers argue, because now many more millions will starve than if famine had struck in the 1960s and we had then been forced to mend our ways.

These bigger issues are outside the scope of the present book. The most relevant point is that the less variety there is among our crops, and the more closely species are tailored by selective breeding to produce optimum yields under optimum conditions, the more vulnerable we become to any change in climatic patterns. And the scale of the problem is indicated most clearly by the extent to which a few varieties now dominate US agriculture, remembering that the US is the breadbasket of the world. By the end of the 1970s only six major varieties of corn covered 71 per cent of the acreage devoted to the crop; just two varieties of peas accounted for 96 per cent of the acreage; 72 per cent of all potato fields were sown with one of four main varieties; and more than half the 28,662 million hectares devoted to wheat was occupied by just 10 varieties. Such a narrow genetic base weeds out lower-yielding varieties, but it also weeds out ruggedness. If all the wheat plants are identical, then they are all vulnerable to the same diseases, the same pests and the same vagaries of weather. A hundred years ago or more, a farmer might grow not just one variety of wheat, but a whole range of crops; if one failed in any year, he had others to fall back on. Today, not just individual farms but great prairies of wheat are sown with the same variety. If anything happens to that particular strain of wheat, the effects ripple around the world.

So the pessimists tell us that we have already gone too far down the wrong road, banking on a continuation of recent climatic patterns and gambling on maximum yields or bust, instead of trying for a guarantee of rather smaller yields whatever happens to the weather. The optimists, on the other hand, would argue that it is not too late to learn from the mistakes of the Green Revolution and set off down a different path. Not only is carbon dioxide itself stimulating to plant growth, but in the regions of the world where it really matters – the poor South – the likely climatic changes could also prove beneficial. Rather than being the last straw to break the back of the world food system, the greenhouse effect may be the key that unlocks the door to a new world free from the threat of famine, and with more equality between North and South.

The optimists

The first cause for optimism about the effects on agriculture of a build-up of carbon dioxide in the atmosphere stems from the direct effect of carbon dioxide on photosynthesis. Plants need carbon dioxide to grow and develop, and under greenhouse conditions enriching the flow of carbon dioxide to plants can produce dramatic improvements in yield. Of course, greenhouse plants are growing under ideal conditions of temperature, nutrient and water supply as well, and in the real world the potential benefits of increased availability of carbon dioxide may be offset by shortage of nutrients or water, or by adverse temperatures. Even so, the effects of carbon dioxide on plants deserve as much publicity as the rest of the saga of the greenhouse effect.

Professor Norman Rosenberg of the Center for Agricultural Meteorology and Climatology at the University of Nebraska–Lincoln, has looked at the way common crop plants respond to higher concentrations of carbon dioxide. Green plants are classified into three main groups according to the way they carry out photosynthesis, and two of these are especially important to agriculture. The so-called C_4 plants include corn, sorghum, millet and sugar cane, while the C_3 plants include wheat, barley, alfalfa, soybean and many others. In practical terms, the main difference between C_3 and C_4 species is that while both of them can carry out respiration by a process independent of the presence of light, C_3 plants have a second respiration system, called photorespiration, which operates only in the daytime. C_4 plants are more efficient at photosynthesis under optimum conditions, but they need strong sunlight and high temperature; C_3 plants are less productive than C_4 plants in the tropics, but above about 45° latitude they continue to thrive reasonably well where the C_4 plants are much less well adapted. Increasing the amount of carbon dioxide available also affects the plants in different ways, producing an increase in photosynthetic activity which is most pronounced in C_3 species, and a decrease in loss of water from the leaves (transpiration) which is most pronounced in C_4 species. Both effects, of course, are just what the farmer ordered.

Very high concentrations of carbon dioxide have detrimental effects on some plants, affecting their development and flowering behaviour. But the greenhouse studies which revealed this involved carbon dioxide concentrations of 2,000 to 5,000 ppm – ten or more times the present atmospheric concentration. At worst, the atmospheric carbon dioxide build-up is unlikely to harm agricultural plants, while at best it may increase yields. The effect is not, in itself, likely to be dramatic, since productivity in most places today is not limited by the amount of carbon dioxide available to the plants, but by the amount of water and nutrients available. But 'if only the direct effects of increased CO_2 concentration in the global atmosphere are considered,' says Rosenberg, 'I speculate, with others, that net photosynthetic production in agricultural crops will increase. Water-use efficiency will be improved in both C_3 and C_4 species.'

One of those 'others' to whom Rosenberg alludes is another eminent agricultural authority, Sylvan Wittwer, who is Director of the Agriculture Experiment Station at Michigan State University.

He has been one of the most outspoken critics of the dire predictions of massive dislocations in agricultural production caused by warming trends and shifts in precipitation patterns. Rising temperatures may be as beneficial as they are bad, he says, and 'the past century provides evidence that US agriculture and its research establishment can cope with and even improve during climatic changes', citing experience of first a warming trend, then a cooling trend, with the Dust Bowl years somewhere in the middle. 'From 1915 to 1945, Indiana farmers experienced a $+0.1°$ C per year trend in temperatures and a total change of $+2°$ C during the past century. American agriculture already has demonstrated that it can adapt to a trend of $+0.1°C$', assuming the year-to-year fluctuations about the trend get no worse.

Looking further afield, Wittwer stresses that global warming would open up new regions of the USSR for crop production, while the US winter wheat zone could *already* be moved 200 miles northwards using new hardy strains that are available today. And looking at the direct effects of carbon dioxide on photosynthesis, Wittwer makes the claim that 'the currently low level of atmospheric carbon dioxide may well be the most limiting factor in overall global agricultural productivity'. In the greenhouse studies he cites, yields peak at a carbon dioxide concentration of 1,000 to 1,200 ppm, with a potential for increased photosynthesis of about 0.5 per cent for each one per cent increase in carbon dioxide concentration for the range up to 300 ppm above present atmospheric levels. According to Wittwer, these effects, which have been known for 20 years or more, represent 'an abandoned gold mine' of agricultural productivity, since astonishingly even today very few greenhouse farmers take the trouble to supply their crops with extra carbon dioxide. As for field-grown crops, 'the higher levels of carbon dioxide now in the earth's atmosphere may already be making a significant contribution to the productivity of agriculture ... a further increase of carbon dioxide in the atmosphere to 400 parts per million (the level projected by some for the year 2020) would probably result in a 20-per-cent increase in photosynthetic rates for some plants, provided other growth factors were not limited'.

This is an extreme view, but one that is quite tenable in the light of present-day agricultural knowledge. Some of the benefits of the Green Revolution may themselves have been due to the increasing concentration of atmospheric carbon dioxide. And, as Wittwer puts it, 'climate change itself should not be viewed exclusively in a negative context. There may be some favorable agricultural outputs' as a result of the shift in temperature and rainfall patterns brought about by the greenhouse effect.

That is certainly the view of Professor Suresh Kumar Sinha, who works at the Indian Agricultural Research Institute in New Delhi. Where Wittwer argues that the USA and the rest of the developed world might benefit from the effect on photosynthesis of a build-up of carbon dioxide, Sinha, speaking at the Stockholm briefing on the greenhouse effect in February 1981, suggested that Third World agriculture may benefit directly from the changing climatic patterns. This echoes, and develops further, some of the often neglected projections made by Bach and other experts from the rich North, but carries more weight coming from an inhabitant of the Third World.

Rice, as well as being a staple for so many people, is an extremely versatile plant – that is why it has become such a widespread staple crop. Rice can be found growing from Czechoslovakia at 49°N to Sumatra on the Equator, and from an altitude of 3,000 metres in Nepal down to sea-level. But it is primarily a crop of the warm, wet regions of the globe, thriving even in Burma, where rainfall reaches 4,500 mm in a year and many other crops would be washed away. The growing season for rice begins and ends when temperatures reach 15°C, and in general a global warming would lengthen the growing season throughout its range, allow more flexibility in plant-

Table 4: The effects of climatic change on rice yields

Rainfall change	Temperature change (°C)					
%	− 2	− 1	− 0·5	+ 0·5	+ 1	+ 2
− 15	− 19	− 13	− 8	− 0·4	0	3
− 10	− 17	− 11	− 6	− 2	2	5
− 5	− 13	− 7	− 2	2	6	9
+ 5	− 9	− 3	2	6	10	13
+ 10	− 5	1	6	10	14	17
+ 15	− 3	3	8	12	16	19

Figures are predicted percentage changes in rice yield. Present world rice production is 300 million tonnes per year. A warmer, wetter climate increases rice yield. (Data from W. Bach, op. cit.)

ing time, and extend the range of rice-growing areas still further. Only a major decrease in rainfall would reduce rice yields, and as Table 4 shows, for a warming of 1° C or more, rice yields would be increased even if rainfall decreased by 15 per cent.

At the Stockholm meeting, the optimism expressed in some quarters was countered by the argument that present-day mono-cultures are so fragile that even increased opportunities to improve productivity may be dangerous. The increased yields Wittwer and others see as possible might only be achieved with greatly increased inputs of nitrate fertilizers, phosphates and other nutrients, while overcropping, the pessimists argue, could simply lead to more dust-bowls in North America. If the North American grain harvest collapsed, the potential for increased productivity in rice-growing regions would count for nothing – famine would strike before im-proved rice yields could be harvested. Professor Sinha's note of cautious optimism was also highly conditional – on fruitful research into water management, the identification of new crop types that could happily tolerate higher temperatures, and the reduction of present dependence on fossil fuels for high crop yields. Even so, he put his finger on the nub of the real carbon dioxide problem by commenting that 'even a remote possibility that CO_2 could in-fluence agriculture favourably in some regions but adversely in some others may have a profound effect on man'.

At this level, carbon dioxide is simply one small part of the world food problem. Many projections foresee a world population of around six billion in the year 2000 – 1·5 billion in developed countries and the remaining 4·5 billion in Asia, Africa and Latin America. Some 2,300 million tonnes of grain would be required to feed such a population at the present level of diet, and the greenhouse effect is likely to have only a marginal influence one way or the other over the next, crucial, twenty years.

Sinha stresses that not all rice varieties like high temperatures, and that in some cases increasing temperatures could cause a significant reduction in yields. So, even at the simplest level, to take advantage of any potentially valuable climatic shifts, farmers in the poorest regions of the world, the most hidebound by tradition, would have to change the habits of generations. A temperature increase of 2° C or more could have severe adverse effects on the varieties of grain now used in India, and Sinha stresses the complexity of the agricul-tural problems involved by pointing out that the optimum temperature for photosynthesis and growth is not necessarily ideal for the development of grains or tubers – the parts of the plant we eat. Amidst all the uncertainties he stressed that 'among all the factors which limit crop yield, water is usually the most important under field conditions'. Regions which until recently produced scarcely anything have become major producers of grain simply because more water is available.

The implication of his research into the problem is that the potential for increased productivity through more efficient agriculture far outweighs the influence of the greenhouse effect, but that there is a great deal of human inertia to be overcome if this is to be achieved. Citing the example of the dry regions of India, he points out that the particular crops people are used to growing 'are not necessarily the best in productivity or income. Crop substitution studies have clearly established that 2 to 4 times more productivity could be obtained from the same land'. But can this be achieved in the real world? The politics of world food suggest that optimism should be tempered with scepticism; but the politics of carbon dioxide may provide the incentive for the effort that is so clearly required.

The politics of food

The world food problem is one of politics and poverty, not an inability of present-day agriculture to grow enough food to feed the world's population. Two or three decades ago, when the United States had grain surpluses and parts of Africa, say, were perceived as inadequately fed, it seemed natural to offer food aid to the less developed countries. But one side-effect of this aid was that local farmers no longer had the incentive to continue with traditional farming, and the agricultural base of many countries collapsed. Now those countries are dependent on the food aid which started out as the icing on their agricultural cake. Coming right up to date, Brazil was until recently self-sufficient in food. The oil price rises of the seventies led, however, to the now famous Brazilian experiment with alcohol fuel for cars. The alcohol comes from sugar cane grown where the Brazilian farmers used to produce food, and the result is that while Brazil's oil imports have dropped, the country has become a net importer of food. In many countries where the survival of most of the population is literally a hand-to-mouth affair, large sums of money are spent on weapons rather than on improving agriculture or providing an industrial base for the economy.*

Sinha's optimism is based more on the prospects of improved agricultural productivity than on any benefits to agriculture related to climatic change, and like everyone who has studied the world food problem he sees climatic change as only a peripheral factor. If we made the most of the world's agricultural potential, any conceivable climatic shifts of the next fifty years would be of no consequence. So much is clear from estimates that the potential agricultural productivity of the world today, using no more than the proven best practice of modern farming more widely applied, is about 50 billion tonnes of grain equivalent per year. India alone could produce $4\frac{1}{2}$ billion tonnes of grain equivalent per year, two thirds of the *world's* minimum requirement to feed the expected 6 billion population in the year 2000.

These figures are startling the first time you see them, and it is hard to take them seriously. They are genuine, but they would require some major shifts in the way the world is run to be achieved, as a couple of examples show. In India, as Sinha points out, farmers are much more efficient in terms of energy use than the farmers of North America. Whereas each tonne of maize produced in India requires the input of energy equivalent to 50 litres of diesel fuel, each tonne produced in the USA involves consumption of 306 litres, six times as much energy. For wheat, the respective figures are 21 and 102 litres per tonne, and for rice 93 and 170 litres per tonne. American farming is already very efficient in terms of yields per acre, and putting more energy in (in the form of fertilizers or mechanization) could only increase production marginally. But, by the same token, cutting down on energy inputs would only reduce yields marginally – and if the energy saved were used by Indian farmers, their yields would increase by much more than the amount needed to cover the American losses.

Of course, that is not the way the world works. But it indicates where the startling figures for maximum possible yields come from, and why the problem is essentially a political one.

* I have discussed all these problems in more detail in my book *Future Worlds*; the Brandt Commission's report *North–South: A Programme for Survival* looks at the state of global affairs today in general terms, while *To Feed This World*, by Sterling Wortman and Ralph Cummings, focuses on the food issue.

The other natural example looks at things on a regional rather than a global scale. If farmers can be persuaded, or educated, to substitute different crops for the ones their forefathers have always grown, the results can be dramatic. In the Indore region of India, for example, substitution of safflower for wheat in 1972 led to a six-fold increase in yields, while substitution of soybean for green gram brought a tripling of yields per hectare. What the Third World still lacks is the application of the same kind of scientific agricultural techniques that transformed farming in the temperate North in the nineteenth century – not the wholesale transfer of inappropriate northern farming methods to the tropics, but the application of the scientific thinking behind those northern techniques to develop their low-latitude equivalents. That plus energy, the recurring fly in the ointment.

In the long term, efficient farming and increased productivity from the land depend upon capital inputs to Third World countries where the potential for improved yields exists. More capital inputs to the already developed farmlands of the rich North are worse than useless, since they divert funds from other regions and can only marginally improve yields. Meanwhile, efficient labour-intensive agriculture, cultivating every corner of the available land, could be a vital stop-gap for many countries, giving a breathing space in which dependence on food imports and food aid could be removed and time provided for long-term investment. If it all sounds like pie in the sky, that is an indictment of the present state of the world, divided as it is into warring camps. The short-term stop-gaps are only that, and can never solve the real problems in the long term. But it is equally an indictment of the way individual governments in the world behave that immediate increases in food production could be achieved in many of the poorest, hungriest regions of the world today by such 'obvious' possibilities as:

1. Maintaining plant cover through the year by growing crops of different ages and different species – a practice used by many 'primitive' farmers but usually dropped hastily in the rush to 'modernize' in the style of the temperate, rich North.

2. Covering bare soil by mulches of leaves, litter, brushwood, or even plastic sheeting, to trap moisture and protect young plants.

3. Using all available nutrient sources – vegetable waste, manure, sewage, compost, ash and so on.

4. Rotating crops to balance nutrients in the soil.

5. Adopting contour ploughing and terracing on sloping lands.

But there is no escaping the fact that food and farming are political issues. In the long term, increasing the wealth of a country never benefits the poor unless the wealth, especially in the form of land, is redistributed, as Keith Griffin describes in his book *Land Concentration and Rural Poverty*. This should not, however, be seen as some Robin Hood gesture which only benefits the poor. It is through an interest in their own land, or communal land, that the poorest are most likely to make the necessary effort to boost production, with resulting benefits throughout the country. Conversely, it is through lack of incentive when working for others, or struggling to raise cash crops to pay the interest on loans, that production is depressed by the present system in so many cases.

In scientific terms, there is no need for anyone in the world to go hungry now or in the foreseeable future. But what conceivable political incentive could there be for the countries of the rich North to provide the kind of help the developing countries need to make progress on the road to agricultural self-sufficiency? With their established track record, the prospect doesn't look too encouraging, which is why the optimists should always be taken with a grain of salt. There is, though, one problem confronting the rich North, and getting more attention each year, which can only be solved by helping Third World countries. Purely in terms of self-interest, it is becoming urgent, in political terms, for the rich North to tackle this problem before it has any chance of getting out of hand, and it can only be tackled with the help of the poor South. The problem is, of course, the carbon dioxide greenhouse effect. And how better to enlist the aid of the countries of the poor South in tackling the problem than by helping them to achieve agricultural self-sufficiency?

The politics of carbon dioxide

If the build-up of carbon dioxide in the atmosphere is seen as a threat to the developed world, and if the computer models are taken at face value, one obvious way to minimize the impact of the resulting climatic changes would be to limit any future production of carbon dioxide so that the concentration in the atmosphere never rose above 50 per cent more than the pre-industrial level. This would almost certainly keep any global increase in temperature below $1°$ C, with few if any detrimental effects on northern agriculture. (If Idso is right, such an increase in carbon dioxide would have no detectable influence on weather patterns at all.) To achieve this aim, a modest increase in fossil fuel consumption could be allowed until the end of this century, but then new technologies would be needed to take over a substantial part of global energy production early in the twenty-first century. There is scarcely any breathing space for the development and introduction of these new technologies, which would have to meet half the world's energy requirements by AD 2010. Even then, such a scenario involves energy growth, worldwide, of less than 3 per cent a year until the new technologies come into the picture, and it seems inconceivable that the Third World would accept a deliberate policy of such low growth rates, which are incompatible with the plans for overall economic development in the poor South.

The developed countries of the North are the ones that will suffer most from the climatic shifts associated with a warmer world, and the main impact of those climatic changes will be on agriculture. The developing nations are, by and large, in those regions of the globe where the temperature and rainfall effects of increased carbon dioxide are likely to be either negligible or advantageous to agriculture in the immediate future. At present the Third World has hardly taken notice of the carbon dioxide 'problem', which pales into insignificance alongside the many more pressing problems it has to tackle. When it does, though, it is easy to imagine the developing countries taking the view that since the build-up of carbon dioxide may actually help their farmers, they should press on with economic development based on exploitation of fossil fuel and leave the rich North to worry.

The greatest concern about carbon dioxide is today expressed in the USA, while Western Europe expresses little sense of urgency about the problem. In the USSR there is some concern, but a general attitude that problems caused by technology can be solved by technology, coupled with a seldom stated but pervasive feeling that in their part of the world a slight warming would be no bad thing. They may be right about Siberia and the prospect of more ice-free northern ports, but they could still face difficulties with grain harvests if regions such as Kazakhstan become drier. Overall, there is really very little chance of keeping carbon dioxide emissions low enough to prevent a rapid doubling of the pre-industrial atmospheric concentration, and rather than waste resources in futile efforts to stem the build-up it makes more sense to accept this as something we have to learn to live with.

The way to minimize the impact of any climatic change on the world is to improve world agriculture, and this could be achieved with a tiny diversion of the capacity of the rich North to constructive help for southern agriculture. The misidentification of the carbon dioxide problem, including scare stories about melting ice-caps, and the possibility of a misdirection of limited resources into futile efforts to tackle the problem by extracting carbon dioxide at great cost and dumping it in the sea, are the worst aspects of the now rolling carbon dioxide bandwagon. The same resources directed to tackling the food problem might begin to get us out from under the threat of weather variability of all kinds.

It may not seem likely that the appropriate effort will be made in the real world. Stanley Ruttenberg of NCAR pointed this out at the 1980 meeting of the American Association for the Advancement of Science:

> Many of the countries that were hard-stricken by the climatic events in the 1970s, and asked for food aid, were net exporters of food. There is no other conclusion except that unavailability of food to some segments of the population was a consequence of national policies. The changing of eating habits of Europeans, Soviets and North Americans, or the placing of fallowed land in the USA back into production, will not solve the problem of those segments of national populations who are the victims of policies, advertent or not, of their own governments and of those who control the agricultural productive system.

In other words, God helps those that help themselves.

Future weather

So the weather is, after all, just one factor in the complex situation facing society today, and it is unlikely to be the dominant factor. We live in interesting times, at the end of an interglacial which, left to its own devices, would return us to a full Ice Age within a few thousand years. The relative warmth of the middle of the twentieth century is, according to many experts using different lines of attack to study the weather machine, probably also over, and, again, if Nature had its way we would be heading back towards colder conditions, returning to the Little Ice Age as a harbinger of the full Ice Age to come. That would be decidedly uncomfortable for the countries of the North, but hardly noticeable in much of the poor South. Against that background, the possibility of a slight global warming thanks to the carbon dioxide greenhouse effect might almost seem a welcome relief, if it were not for the way the highly productive agricultural regions of the North are so finely tuned to 'normal' weather. But there are no certainties in science, only probabilities, and the only safe prediction we can make about the weather of the next fifty years is that it will probably be different, but not very different, from the weather of the past fifty years. The odds are far too poor to justify a gamble on either warming or cooling, and the only sensible option is to hedge our bets with investment not in more fine-tuning of agriculture to produce even greater yields if the weather is just right, and nothing if the weather is just wrong, but in developing more rugged reliability to ensure that most of the crop is good all of the time.

The puzzle of future weather is fascinating, but it is nothing compared with the real problems facing the world today, and it would be a mistake to divert too much attention on to the possibilities of either a greenhouse Earth or a Little Ice Age, and away from those real, and urgent, problems of development, food and energy. The tragedy will be if the 'threat' of the greenhouse effect leads to an ill-conceived and expensive search for a technofix, gobbling up resources that could better be used to improve world agriculture and reduce our susceptibility to weather variations. That is the real carbon dioxide 'problem' facing mankind today.

Bibliography

Aagard, K., and Coachman, L. K. 'Toward an ice-free Arctic Ocean', *EOS*, vol. 56 (1974), p. 484.

Ackerman, M., Lippens, C., and Lechevallier, M. 'Volcanic material from Mount St Helens in the stratosphere over Europe', *Nature*, vol. 287 (1980), p. 614.

Albanese, A. S., and Steinberg, M. 'Environmental control technology for atmospheric carbon dioxide', published as Report 006 in the US Department of Energy Carbon Dioxide Effects Research and Assessment Program, US Department of Energy, Washington DC (1980).

Angione, R., Medeiros, E., and Roosen, R. 'Stratospheric ozone as viewed from the Chappuis band', *Nature*, vol. 261 (1976), p. 289.

Anonymous. 'Two-way split generates hydrogen more efficiently', *New Scientist*, vol. 92 (8 October 1981), p. 100.

Arrhenius, S. 'On the influence of carbonic acid in the air upon the temperature of the ground', *Philosophical Magazine*, vol. 41 (1896), p. 237.

Bach, W. 'The potential consequences of increasing CO_2 levels in the atmosphere', in *Carbon Dioxide, Climate and Society*, ed. J. Williams, Pergamon, Oxford (1978).

Baldwin, B., Pollack, J., Summers, A., Toon, O., Sagan, C., and van Camp, W. 'Stratospheric aerosols and climatic change', *Nature*, vol. 263 (1976), p. 551.

Barnett, T. P. 'The role of the Ocean in the global climate system', in *Climatic Change*, ed. J. Gribbin, Cambridge University Press, London and New York (1978).

Bernard, H. W. *The Greenhouse Effect*, Ballinger, Cambridge, Massachusetts (1980).

Bolin, B. 'Changes of land biota and their importance for the carbon cycle', *Science*, vol. 183 (1977), p. 613.

Brandt Commission. *North–South: A Programme for Survival*, Pan, London (1980).

Broecker, W. S. 'Climatic change: Are we on the brink of a pronounced global warming?', *Science*, vol. 189 (1975), p. 460.

Bryson, R. A., and Dittberner, G. J. 'A non-equilibrium model of hemispheric mean surface temperature', *Journal of the Atmospheric Sciences*, vol. 33 (1976), p. 2094.

Bryson, R. A., and Murray, T. J. *Climates of Hunger*, University of Wisconsin Press (1977).

Bucha, V., Taylor, R. E., Berger, R., and Haury, E. W. 'Geomagnetic intensity: changes during the last 9000 years in the western hemisphere', *Science*, vol. 168 (1970), p. 111.

Buringh, P., Van Heemst, H. D., and Staring, G. J. *Computation of the Absolute Maximum Food Production of the World*, Agricultural University, University of Wageningen, Netherlands (1975).

Callis, L. B., Natarajan, M., and Nealey, J. E. 'Ozone and temperature trends associated with the 11-year solar cycle', *Science*, vol. 204 (1979), p. 1303.

Coakley, J. A. Study reported by Ralph Segman of NCAR Information Office in release headed 'Volcanic dust has greater impact on climate than expected' (7 December 1979).

Council of Environmental Quality (US). *Global Energy Futures and the Carbon Dioxide Problem*, US Government Printing Office, Washington D.C. (1981).

Crutzen, P. J., Isaksen, I. S. A., and Reid, G. C. 'Solar proton events: stratospheric sources of nitrogen oxide', *Science*, vol. 189 (1975), p. 457.

Cunningham, C. 'A world full of coal', *New Scientist*, vol. 90 (7 May 1981), p. 338.

Currie, R. G. 'Distribution of solar cycle signal in surface air temperature over North America', *Journal of Geophysical Research*, vol. 84 (1979), p. 753.

Currie, R. G. 'Solar cycle signal in air temperature in North America: amplitude, gradient, phase and distribution', *Journal of the Atmospheric Sciences*, vol. 38 (1981), p. 808.

Dansgaard, W., Johnson, S. J., Reeh, N., Gundestrup, N., Clausen, H. B., and Hammer, C. U. 'Climatic changes, Norsemen, and modern man', *Nature*, vol. 255 (1975), p. 24.

Delmas, R. J., Ascencio, J.-M., and Legrand, M. 'Polar ice evidence that atmospheric CO_2 20,000 years B.C. was 50% of present?', *Nature*, vol. 284 (1980), p. 155.

Department of Energy (US). 'Workshop on the global effect of carbon dioxide from fossil fuels', Report 001 in the US Department of Energy Carbon Dioxide Effects Research and Assessment Program, US Department of Energy, Washington D.C. (1979). This is the first in a series of Department of Energy reports on the carbon dioxide problem; a comprehensive interim report is planned for 1984, and a final report on a ten-year research programme in 1989. By April 1981 16 reports had been published. Available from the US Department of Commerce, Springfield, Virginia.

Donn, W. L., and Shaw, D. M. 'Model of climate evolution based on continental drift and polar wandering', *Geological Society of America Bulletin*, vol. 88 (1977), p. 390.

Ellis J. S., Von de Haar, T. H., Levitus, S., and Oort, A. H. 'The annual variation in the global heat balance of the earth', *Journal of Geophysical Research*, vol. 83 (1978), p. 1958.

Epstein, L. C., and Hewitt, P. G. *Thinking Physics*, Insight Press, San Francisco (1981).

Flohn, H. 'Climate and energy: a scenario to a 21st-century problem', *Climatic Change*, vol. 1 (1977), p. 5.

Foukal, P. V., Mack, P. E., and Vernazza, J. E. 'The effects of sunspots and faculae on the solar constant', *Astrophysical Journal*, vol. 215 (1977), p. 952.

Francis, P. *Volcanoes*, Penguin, Harmondsworth (1976).

Franklin, B. *The Writings of Benjamin Franklin*, ed. A. H. Smyth, vol. IX (1783–8), pp. 217–18, Haskell House, New York (1970).

Freeman, C., and Jahoda, M. (eds.). *World Futures: The Great Debate*, Martin Robertson, London/Universe, New York (1978).

Gribbin, J. *Climate and Mankind*, Earthscan, London (1979); and *Carbon Dioxide, the Climate, and Man*, Earthscan, London (1981).

Gribbin, J. *The Death of the Sun*, Delta, New York (1981).

Gribbin, J. *Future Worlds*, Abacus, London/Plenum, New York (1981).

Gribbin, J., and Lamb, H. H. 'Climatic change in historical times', in *Climatic Change*, ed. J. Gribbin, Cambridge University Press, London and New York (1978).

Griffin, K. *Land Concentration and Rural Poverty*, Macmillan, London (1975).

Hammer, C. V., Clausen, H. B., and Dansgaard, W. 'Greenland ice-sheet evidence of post-glacial volcanism and its climatic impact', *Nature*, vol. 288 (1980), p. 230.

Hansen, J. E., Wang, W.-C., and Lacis, A. A. 'Mount Agung eruption provides test of a global climatic perturbation', *Science*, vol. 199 (1978), p. 1065.

Hansen, J., Johnson, D., Lacis, A., Lebedeff, S., Lee, P., Rind, D., and Russell, G. 'Climatic import of increasing atmospheric carbon dioxide', *Science*, vol. 213 (1981), p. 957.

Hays, J. D., Imbrie, J., and Shackleton, N. J. 'Variations in the Earth's orbit: Pacemaker of the Ice Ages', *Science*, vol. 194 (1976), p. 1121.

Heath, D. F., Krueger, A. J., and Crutzen, P. J. 'Solar proton event: influence on stratospheric ozone', *Science*, vol. 197 (1977), p. 886.

Henderson-Sellen, A., and Schwarz, A. W. 'Chemical evolution and ammonia in the early Earth's atmosphere', *Nature*, vol. 287 (1980), p. 526.

Hoyle, F. *Ice*, Hutchinson, London (1981).

Idso, S. B. 'The climatological significance of a doubling of Earth's atmospheric carbon dioxide concentration', *Science*, vol. 207 (1980), p. 1462.

Imbrie, J., and Imbrie, K. *Ice Ages: Solving the Mystery*, Macmillan, London/Enslow, New York (1979).

Kandel, R. S. 'Surface temperature sensitivity to increased atmospheric CO_2', *Nature*, vol. 293 (1981), p. 634.

Kaufmann, W. J. *Planets and Moons*, Freeman, San Francisco (1979).

Kellogg, W. W. 'Global influences of mankind on the climate', in *Climatic Change*, ed. J. Gribbin, Cambridge University Press, London and New York (1978).

Kellogg, W. W., and Schware, R. *Climatic Change and Society*, Aspen Institute/Westview Press (1981).

Kelly, P. M. 'Volcanic dust veils and North Atlantic climate change', *Nature*, vol. 268 (1977), p. 616.

Kelly, P. M., and Lamb, H. H. 'Prediction of volcanic activity and climate', *Nature*, vol. 262 (1976), p. 5.

King, J. W. 'Weather and the Earth's magnetic field', *Nature*, vol. 247 (1974), p. 131.

Kerr, R. A. 'Carbon dioxide and climate: carbon budget still unbalanced', *Science*, vol. 197 (1977), p. 1352.

Kerr, R. A. 'Carbon budget not so out of whack', *Science*, vol. 208 (1980), p. 1358.

Kondratyev, K. Y. 'The "Greenhouse Effect" of minor constituents in the atmosphere', *Weather*, vol. 35 (September 1980), p. 252.

Kondratyev, K. Y., and Nikolsky, G. A. 'Solar radiation and solar activity,' *Quarterly Journal of the Royal Meteorological Society*, vol. 96 (1970), p. 509.

Kondratyev, K. Y., and Nikolsky, G. A. 'The stratospheric mechanism of solar and anthropogenic influences on climate', in *Solar Terrestrial Influences on Weather and Climate*, ed. B. M. McCormac and T. A. Seliga, Reidel, Dordrecht (1979).

Kukla, G., Berger, A., Lotti, R., and Brown, J. 'Orbital signature of interglacials', *Nature*, vol. 290 (1981), p. 295.

Kutzbach, J. E. 'Monsoon climate of the early Holocene: climate experiment with the Earth's orbital parameters for 9,000 years ago', *Science*, vol. 214 (1981), p. 59.

Lamb, H. H. 'Volcanic dust in the atmosphere; with a chronology and assessment of its meteorological significance', *Philosophical Transactions of the Royal Society*, vol. A266 (1970), p. 425.

Lamb, H. H. *Climate: Present, Past and Future*, vol. 1, Methuen, London (1972); vol. 2, Methuen, London (1977).

Lamb, H. H. 'Geomagnetism and climatic change', *Nature*, vol. 247 (1974), p. 127.

Leach, G., Lewis, C., Romig, F., Foley, G., and Van Buren, A. *A Low Energy Strategy for the United Kingdom*, Science Reviews/ I.I.E.D., London (1979).

Libby, L. M., Pandolfi, L. J., Pryton, P. H., Marshall, J., Becker, B., and Giertz-Sienbenlist, V. 'Isotopic tree thermometers', *Nature*, vol. 261 (1976), p. 284.

Lockwood, J. G. *Causes of Climate*, Edward Arnold, London (1979).

McDonald, A. *Energy in a Finite World*, Executive summary of a study by International Institute for Applied Systems Analysis, IIASA, Laxenburg, Austria (1981).

Madden, R. A., and Ramanathan, V. 'Detecting climate change due to increasing carbon dioxide', *Science*, vol. 209 (1980), p. 763.

Manabe, S., and Wetherald, R. T. 'Thermal equilibrium of the atmosphere with a given distribution of relative humidity', *Journal of Atmospheric Sciences*, vol. 24 (1967), p. 241.

Manabe, S., and Wetherald, R. T. 'The effects of doubling the CO_2 concentration on the climate of a general circulation model', *Journal of Atmospheric Sciences*, vol. 32 (1975), p. 3.

Manabe, S., and Wetherald, R. T. 'On the distribution of climate change resulting from an increase in the CO_2 content of the atmosphere', *Journal of Atmospheric Sciences*, vol. 37 (1980), p. 99.

Marchetti, C. 'On geoengineering and the CO_2 problem', *Research Memorandum, RM-76-17*, International Institute for Applied Systems Analysis, Vienna (1976); see also *Climatic Change*, vol. 1 (1977), p. 59.

Mercer, J. H. 'West Antarctic ice sheet and CO_2 greenhouse effect: a threat of disaster', *Nature*, vol. 271 (1978), p. 321.

Miller, A. A., and Parry, M. *Everyday Meteorology*, Hutchinson, London (revised edition, 1975).

Murray Mitchell, Jr, J. 'A reassessment of atmospheric pollution as a cause of long-term changes of global temperature', in *The Changing Global Environment*, ed. S. Fred Singer, Reidel, Dordrecht (1975).

Mustacchi, C., Armenante P., and Cena, V. 'Carbon dioxide disposal in the ocean', in *Carbon Dioxide, Climate and Society*, ed. J. Williams, Pergamon Press, London (1978).

Myers, N. 'The Hamburger Connection', *Ambio*, vol. 10 (1981), p. 3.

Namias, J. 'Some concomitant regional anomalies associated with hemispherically averaged temperature variations', *Journal of Geophysical Research*, vol. 85 (1980), p. 1585.

National Research Council (US). *Energy and Climate*, NAS, Washington D.C. (1977).

Newell, R. E., and Dopplick, T. G. 'Questions concerning the possible influence of anthropogenic CO_2 on atmospheric temperature', *Journal of Applied Meteorology*, vol. 18 (1979), p. 822.

Ninkovich, D., and Donn, W. L. 'Explosive Cenozoic volcanism and climatic implications', *Science*, vol. 194 (1976), p. 899.

Ojakangas, R. W., and Darby, D. G. *The Earth: Past and Present*, McGraw-Hill, New York (1976).

Öpik, E. J. *Climate and the Changing Sun*, Scientific American reprint No. 835, W. H. Freeman, San Francisco (June 1958).

Parry, M. *Climatic Change, Agriculture and Settlement*, Wm. Dawson, Kent & Archon Books, Connecticut (1978).

Porter, S. C. 'Recent glacier variations on volcanic eruption', *Nature*, vol. 291 (1981), p. 139.

Reck, R. A. 'Aerosols in the atmosphere: calculation of the critical absorption/backscatter ratio', *Science*, vol. 186 (1974), p. 1034.

Reck, R. A. 'Aerosols and polar temperature changes', *Science*, vol. 188 (1975), p. 728.

Reid, G. C., Isaksen, I. S. A., Holger, T. E., and Crutzen, P. J. 'Influence of ancient solar proton events on the evolution of life', *Nature*, vol. 259 (1976), p. 177.

Revelle, R. R., and Shapero, D. C. 'Energy and climate', *Environmental Conservation*, vol. 5 (1978), p. 81.

Roberts, W. O., and Lansford, H. *The Climate Mandate*, W. H. Freeman, San Francisco (1979).

Robock, A. 'The "Little Ice-Age": northern hemisphere average observations and model calculations', *Science*, vol. 206 (1979), p. 1402.

Robock, A. 'The Mount St Helens volcanic eruption of 18 May 1980: minimal climatic effect', *Science*, vol. 212 (1981), p. 1383.

Rosenberg, N. J. 'The increasing CO_2 concentration in the atmosphere and its implication on agricultural productivity', *Climatic Change*, vol. 3 (1981), p. 265.

Rotty, R. M. 'The atmospheric CO_2 consequences of heavy dependence on coal', in *Carbon Dioxide, Climate and Society*, ed. Jill Williams, Pergamon Press, London (1978).

Ruddiman, W. F., and McIntyre, A. 'Oceanic mechanisms for amplification of the 23,000-year-ice-volume cycle', *Science*, vol. 212 (1981), p. 617.

Ruttenberg, S. 'Inclusion of societal factors in assessing the impact of climate on food supplies', paper presented at the San Francisco meeting of the American Association for the Advancement of Science (January 1980).

Schmitt, L. E. (ed.). 'Proceedings of the Carbon Dioxide and Climate Research Program Conference on April 24–25, 1980', Report 011 in the US Department of Energy Carbon Dioxide Effects Research and Assessment Program, US Department of Energy, Washington D.C. (1980).

Schneider, S. H. 'On the carbon dioxide–climate confusion', *Journal of Atmospheric Sciences*, vol. 32 (1975), p. 2060.

Schneider, S. H. 'The trillion dollar question', *Aware*, No. 114 (March 1980), p. 10.

Schneider, S. H., and Chen, R. S. 'Carbon dioxide warming and coastline flooding: a problem review and exploratory climatic impact assessment', *Annual Review of Energy*, vol. 5 (1980), p. 107.

Schneider, S. H., Kellogg, W. W., and Ramanathan, V. 'Carbon Dioxide', *Science*, vol. 210 (1980), p. 6.

Schneider, S. H., and Mass, C. 'Volcanic dust, sunspots and temperature trends', *Science*, vol. 190 (1975), p. 741.

Sear, C. B., and Lough, J. M. 'Three scenarios for a warm, high CO_2 world', paper presented to the American Meteorological Society, First Conference on Climatic Variations, San Diego, California (January 1981).

Sherwood, M. 'Full Speed Ahead on artificial water splitting', *New Scientist*, vol. 88 (1980), p. 504.

Siegenthaler, V., and Oeschger, H. 'Predicting future atmospheric carbon dioxide levels', *Science*, vol. 199 (1978), p. 388.

Sinha, S. K., and Swaminathan, M. S. 'The absolute maximum food production in India – an estimate', *Current Science*, vol. 48 (1979), p. 425.

Smith, P. J. 'Does the geomagnetic field affect climate?', *Nature*, vol. 249 (1974), p. 511.

Stommel, H., and Stommel, E. 'The year without a summer', *Scientific American*, vol. 240 (June 1979), p. 134.

Stuiver, M. 'On climatic changes', *Quarternary Research*, vol. 2 (1972), p. 409.

Stuiver, M. 'Atmospheric carbon dioxide and carbon reservoir changes', *Science*, vol. 199 (1978), p. 253.

Sunspot Index Data Centre, Brussels. *Sunspot Bulletin* (monthly).

Thompson, S. L., and Schneider, S. H. 'Carbon dioxide and climate: ice and ocean', *Nature*, vol. 290 (1981), p. 9.

Wang, W.-C., Yung, Y. L., Lacis, A. A., Mo, T., and Hansen, J. E. 'Greenhouse effects due to man-made perturbations of trace gases', *Science*, vol. 194 (1976), p. 685.

Wetherald, R. T., and Manabe, S. 'Cloud cover and climate sensitivity', *Journal of the Atmospheric Sciences*, vol. 37 (1980), p. 1485.

Wexler, H. 'Volcanoes and world climate', *Scientific American*, vol. 186 (April 1952), p. 74.

Wigley, T. M. L., and Brimblecombe, P. 'Carbon dioxide, ammonia, and the origin of life', *Nature*, vol. 291 (1981), p. 713.

Wigley, T. M. L., and Jones, P. D. 'Detecting CO_2 induced climatic change', *Nature*, vol. 292 (1981), p. 205.

Wigley, T. M. L., Jones, P. D., and Kelly, P. M. 'Scenario for a warm, high-CO_2 world', *Nature*, vol. 283 (1980), p. 17.

Willett, H. C. 'The sun as maker of weather and climate', *Technology Review* (January 1976), p. 47.

Willett, H. C. 'The solar prediction of climatic changes', paper presented to a workshop on solar–terrestrial prediction, held at Boulder, Colorado (21 April 1979).

Williams, J. 'Global energy strategies: the implication of CO_2', *Futures*, vol. 10 (1978), p. 293.

Williams, J. (ed.). *Carbon Dioxide, Climate and Society: IIASA Workshop Proceedings, 1978*, Pergamon, Oxford (1978).

Williams, J. 'Anomalies in temperature and rainfall during warm Arctic seasons as a guide to the formulation of climate scenarios', *Climatic Change*, vol. 12 (1980), p. 249.

Wilson, A. T. 'Pioneer agriculture explosion and CO_2 levels in the atmosphere', *Nature*, vol. 273 (1978), p. 40.

de Witte, H. 'Research into cheaper methods of manufacturing solar cells', *Science Policy in the Netherlands*, vol. 3 (June 1981), p. 14.

Wittwer, S. H. 'Carbon dioxide and climatic change: an agricultural perspective', *Journal of Soil and Water Conservation*, vol. 35 (1980), p. 116.

Wollin, G., Ericson, D. B., Ryan, W. B. F., and Foster, J. H. 'Magnetism of the Earth and climatic changes', in *Earth and Planetary Science Letters*, vol. 12 (1971), p. 175.

Wollin, G., Ericson, D. B., and Ryan, W. B. F. 'Variations in magnetic intensity and climate changes', *Nature*, vol. 232 (1971), p. 549.

Wollin, G., Ryan, W. B. F., and Ericson, D. B. 'Climatic changes, magnetic intensity, variations and fluctuations of the eccentricity of the Earth's orbit during the past 2,000,000 years and a mechanism which may be responsible for the relationship', in *Earth and Planetary Science Letters*, vol. 41 (1978), p. 395.

Woodwell, G. M. 'The carbon dioxide question', *Scientific American*, vol. 283 (January 1978), p. 34.

Woodwell, G. M., Whittaker, R. H., Reiners, W. A. Likens, G. E., Delwiche, C. C., and Botkin, D. B. 'The biota and the world carbon budget', *Science*, vol. 199 (1978), p. 141.

Wortman, S., and Cummings, R. *To Feed This World*, Johns Hopkins University Press, Baltimore, Maryland (1978).

Index

Page numbers in italics refer to Figures. Tables are indicated by 't' after the page number.